August :

AF822674

Gerald S Berenson

Evolution of Cardio-Metabolic Risk from Birth to Middle Age

Gerald S. Berenson
Editor

Evolution of Cardio-Metabolic Risk from Birth to Middle Age

The Bogalusa Heart Study

Springer

Editor
Dr. Gerald S. Berenson
Department of Medicine,
Pediatrics, Biochemistry, Epidemiology
Tulane University School of Medicine and School
of Public Health and Tropical Medicine
Canal Street 1440, 70112 New Orleans, LA
USA
berenson@tulane.edu

ISBN 978-94-007-1450-2 e-ISBN 978-94-007-1451-9
DOI 10.1007/978-94-007-1451-9
Springer Dordrecht Heidelberg London New York

Library of Congress Control Number: 2011930402

© Springer Science+Business Media B.V. 2011
No part of this work may be reproduced, stored in a retrieval system, or transmitted in any form or by any means, electronic, mechanical, photocopying, microfilming, recording or otherwise, without written permission from the Publisher, with the exception of any material supplied specifically for the purpose of being entered and executed on a computer system, for exclusive use by the purchaser of the work.

Printed on acid-free paper

Springer is part of Springer Science+Business Media (www.springer.com)

Foreword

The Bogalusa Study and the work of Dr. Gerald Berenson and colleagues is one of the great successes of recent cardiovascular research. In the early 1970s, the late Dr. Al Tyroler and I had the opportunity of reviewing and site visiting the original grant proposal for a study of cardiovascular risk factors in black and white children in a small community, Bogalusa LA, outside of New Orleans. The Principal Investigator was a highly regarded cardiologist with a major interest in mucopolysaccharides and atherosclerosis. His family had, for a long time, very close ties to the Bogalusa community. Dr. Berenson recognized the potential of the Bogalusa community to provide a scientific base for the origins of atherosclerosis and cardiovascular disease (CVD) beginning in childhood. The National Heart, Lung, and Blood Institute (NHLBI) had recognized that the study of early origins of CVD beginning in childhood was an important component of their research portfolio. It took little effort to convince an external review group that Dr. Berenson and his team at the LSU School of Medicine could do the study. The Bogalusa Study is clearly one of the very best investments that the NHLBI made back in the early 1970s. Dr. Berenson and I still discuss the early advisory committee meetings when we discussed how to develop a series of hypotheses to test in the Bogalusa Study. He was very fortunate in the early years to have an outstanding statistician, CA MacMahon, working with him at LSU and the late epidemiologist, Antonie Voors.

This book reflects the important and unique aspects of the Bogalusa Study: (1) The ability to maintain a cohort from childhood to adult life with an adequate sample size in a defined population to compare black-white differences in determinants of risk factors; (2) the ability to combine excellent and modern physiological and biochemical methodology to an epidemiology population study, i.e. carotid intima media thickness, pulse wave velocity, telomere length, heart rate variability, pathology of coronary arteries; and (3) a very strong commitment to prevention of CVD and especially primordial prevention of key risk factors using community resources, especially the education system.

The book reflects all 3 of these very successful components of the Bogalusa Study and especially the importance of Dr. Berenson. His message of the need for prevention of risk factors beginning in childhood is supported by the overwhelming evidence that has been generated initially from the Bogalusa Study and then further supported by similar research in many countries, as also reported in this book.

Dr. Berenson is a pioneer in demonstrating that excellent clinical cardiovascular methods could be applied to a study of the evaluation of risk factors and pathology in children and adults. The study population continues to provide an extraordinary resource for epidemiology research. The biggest reward from the Bogalusa Study is whether we can translate the great successes of this study to prevention of CVD beginning early in life as articulated throughout this book, especially in Chaps 11 and 14.

Distinguished University
Professor of Public Health

Lewis H. Kuller, MD DrPH

Preface

It is a pleasure to highlight some of the findings over the past decade in this third book from the Bogalusa Heart Study. Tribute and appreciation has to be given to the Study subjects, some of whom have participated since 1973. I am overwhelmed with gratitude by the support from the community of Bogalusa and high level of participation that enabled us to obtain perspectives on the evolution of cardiovascular risk from birth through mid-adulthood. This Study is unique in that it remains the only long term study in a biracial (black/white) population beginning in childhood.

Recognition needs to be given to the many investigators and support staff that have helped conduct the study: They have exemplified the highest level of commitment and devotion to make this Study successful. Also funding from the National Heart Lung and Blood Institute (NHLBI), National Institute on Aging (NIA), National Institute on Child Health and Human Development (NICHD) and the American Heart Association (AHA), was crucial, without which the Study could not have been conducted. Such support made it possible to unravel as much about the natural history of the early origin of coronary artery disease, essential hypertension and type II diabetes mellitus, as our team did.

The essence of the many publications of this Study all clearly indicate coronary artery disease, as a prelude to coronary heart disease, primary hypertension, and diabetes, all have their origin in childhood, even with evidence to begin in utero (website: http://tulane.edu/som/cardiohealth/index.cfm). We have noted the importance of a strong family history, that will become more evident as genetic studies evolve. We found risk factors can be developed in early life to diagnose these conditions and show their "silent" burden on the cardiovascular system beginning in childhood. In fact, discussions to follow consider fetal origins of risk factors set the stage for observations beginning in childhood and the need for primordial prevention. The studies of chromosomal telomeres provide some insight into the aging process and impact of environmental stress reflecting black-white contrasts on cardiovascular diseases and the ethnic and gender variations of morbidity and mortality in our population. Understanding such variations provide a background to aid both clinical management and approaches to prevention. A theme throughout the chapters use race and gender contrasts to reflect on different mechanisms and complexity of factors related to development of cardiovascular diseases beginning from early childhood.

It has been rewarding to me to have founded this program with my colleagues who helped to begin with laboratory and experimental studies related to the complex sugars and connective tissues of arterial wall in atherosclerosis, and to be encouraged by findings from my colleagues in Pathology who earlier found the presence of atherosclerotic lesions in childhood. My interest in cardiovascular disease and the concept of meaningful risk factors related to heart disease in adults by the Framingham Study, set the stage to begin studying children at a population level. Suggestions from my many colleagues have been invaluable.

As a corollary to the Bogalusa Heart Study, effective prevention programs have been developed from the findings based on lifestyles and behavior learned in Bogalusa. Application of health education for children in the general public can help abort or at least delay the cardiovascular maladies so common in our society and world-wide. It is our hope that the potential from prevention beginning in childhood will become recognized as an acceptable and common practice. This is our way to address quality of life from its origin and maybe extend to the end of life.

Gerald S. Berenson, MD

Contents

Contributors

Abraham Aviv The Center of Human Development and Aging, New Jersey Medical School, University of Medicine and Dentistry of New Jersey, Newark, NJ, USA
e-mail: avivab@umdnj.edu

Gerald S. Berenson, MD Department of Medicine, Pediatrics, Biochemistry, Epidemiology, Center for Cardiovascular Health, Tulane University School of Medicine and School of Public Health and Tropical Medicine, New Orleans, LA, USA
e-mail: berenson@tulane.edu

Heidi M. Blanck, PhD Division of Nutrition, Physical Activity and Obesity, Obesity Prevention and Control Branch, Centers for Disease Control and Prevention K-26, Atlanta, GA, USA
e-mail: hblanck@cdc.gov

Wei Chen, MD PhD Department of Epidemiology, Center for Cardiovascular Health, Tulane University School of Public Health and Tropical Medicine, New Orleans, LA, USA
e-mail: wchen1@tulane.edu

Pronabesh DasMahapatra, MD MPH Department of Epidemiology, Center for Cardiovascular Health, Tulane University School of Public Health and Tropical Medicine, New Orleans, LA, USA
e-mail: pdasmaha@tulane.edu, pronabesh@yahoo.com

James Ebert, MD, MBA, MPH, FAAP Boonshoft School of Medicine, Wright State University, Dayton, OH, USA
e-mail: james.ebert@wright.edu

Camilo Fernandez Alonso, MD MS Department of Epidemiology, Center for Cardiovascular Health, Tulane University School of Public Health and Tropical Medicine, New Orleans, LA, USA
e-mail: cfernan1@tulane.edu, camilo@ginfesa.com

David S. Freedman, PhD Division of Nutrition, Physical Activity and Obesity, Obesity Prevention and Control Branch, Centers for Disease Control and Prevention K-26, Atlanta, GA, USA
e-mail: dxf1@cdc.gov

Thomas Giles, MD Department of Medicine, Heart and Vascular Institute, Tulane University School of Medicine, New Orleans, LA, USA
e-mail: tgiles1@tulane.edu

Markus Juonala, MD PhD Department of Pediatrics, Medicine and Clinical Physiology, Research Center of Applied and Preventive Cardiovascular Medicine, University of Turku and Turku University Hospital, Turku, Finland
e-mail: mataju@utu.fi

Costan G. Magnussen, PhD Department of Pediatrics, Medicine and Clinical Physiology, Research Center of Applied and Preventive Cardiovascular Medicine, University of Turku and Turku University Hospital, Turku, Finland
e-mail: Costan.Magnussen@utas.edu.au

Murdoch Childrens Research Institute, Melborne, Australia
e-mail: costan.magnussen@utu.fi

Quoc Manh Nguyen, MD MPH Department of Epidemiology, Center for Cardiovascular Health, Tulane University School of Public Health and Tropical Medicine, New Orleans, LA, USA
e-mail: qnguyen1@tulane.edu

Theresa A. Nicklas, DrPH Department of Pediatrics, Baylor College of Medicine, Houston, TX, USA
e-mail: tnicklas@bcm.edu

Harri Niinikoski, MD PhD Department of Pediatrics, Medicine and Clinical Physiology, Research Center of Applied and Preventive Cardiovascular Medicine, University of Turku and Turku University Hospital, Turku, Finland
e-mail: harri.niinikoski@tyks.fi

Carol E. O'Neil, PhD RD Human Nutrition and Food, School of Human Ecology, Louisiana State University, Baton Rouge, LA, USA
e-mail: coneil1@lsu.edu

Marietta Orlowski, PhD Department of Community Health, Boonshoft School of Medicine, Wright State University, Dayton, OH, USA
e-mail: marietta.orlowski@wright.edu

Sandra Owen, BSN, MEd, FASHSA Emerita, College of Education, Georgia State University, Atlanta, USA
e-mail: sowen2@gsu.edu

Arthur Pickoff, MD Pediatrics and Community Health, Boonshoft School of Medicine, Wright State University, Dayton, OH, USA
e-mail: arthur.pickoff@wright.edu

Olli T. Raitakari, MD PhD Department of Pediatrics, Medicine and Clinical Physiology, Research Center of Applied and Preventive Cardiovascular Medicine, University of Turku and Turku University Hospital, Turku, Finland
e-mail: olli.raitakari@utu.fi

Olli Simell, MD PhD Department of Pediatrics, Medicine and Clinical Physiology, Research Center of Applied and Preventive Cardiovascular Medicine, University of Turku and Turku University Hospital, Turku, Finland
e-mail: olli.simell@utu.fi

Sathanur R. Srinivasan, PhD Department of Epidemiology, Biochemistry, Center for Cardiovascular Health, Tulane University School of Public Health and Tropical Medicine, New Orleans, LA, USA
e-mail: ssriniv1@tulane.edu

Elaine M. Urbina, MD Department of Pediatrics, Division of Endocrinology, College of Medicine, University of Cincinnati, Cincinnati, OH, USA
e-mail: Elaine.Urbina@cchmc.org

Preventive Cardiology, Cincinnati Children's Hospital Medical Center, Cincinnati, Ohio, USA

Jorma S.A. Viikari, MD PhD Department of Pediatrics, Medicine and Clinical Physiology, Research Center of Applied and Preventive Cardiovascular Medicine, University of Turku and Turku University Hospital, Turku, Finland
e-mail: jorma.viikari@utu.fi

Jihua Xu, MD Department of Epidemiology, Center for Cardiovascular Health, Tulane University School of Public Health and Tropical Medicine, New Orleans, LA, USA
e-mail: jxu@tulane.edu

Chapter 1
Exploring Chromosomal Leukocyte Telomere Length Dynamics in the Bogalusa Heart Study

Abraham Aviv and Wei Chen

Abstract Leukocyte telomere length (LTL) is a biomarker of human aging in that it is relatively short in individuals who display aging-related diseases, principally atherosclerosis. The Bogalusa Heart Study (BHS) has provided unique settings to explore the mechanisms that impact LTL dynamics (LTL and its age-dependent attrition) in young adults. This chapter briefly reviews the background of LTL research and original observations on LTL dynamics and the relations to various indices of cardiovascular aging in the black–white cohort of the BHS. Specifically, the results based on both cross-sectional and longitudinal analyses, black–white difference, and genetic study are summarized. By now, there is a vast and sometimes conflicting literature about the links of LTL with aging and aging-related diseases. The original observations in the BHS were instrumental for the development of a whole new look at what LTL dynamics are all about and in what way they enlighten us about human aging.

Keywords Telomere length • Cardiovascular aging • Black–white difference • Longitudinal analysis

1.1 Introduction

The Bogalusa Heart Study (BHS) has provided unique settings to explore the mechanisms that impact leukocyte telomere length (LTL) dynamics (LTL and its age-dependent attrition) in young adults. The Study is distinguished not necessarily by its longitudinal nature; major studies such as the Framingham Heart Study and the Cardiovascular Health Study have also followed their participants for many years. However, BHS participants were recruited during childhood several decades ago and most have been followed up ever since. This feature, the biracial composition

A. Aviv (✉)
The Center of Human Development and Aging, New Jersey Medical School,
University of Medicine and Dentistry of New Jersey, Newark, NJ, USA
e-mail: avivab@umdnj.edu

G. S. Berenson (ed.), *Evolution of Cardio-Metabolic Risk from Birth to Middle Age*,
DOI 10.1007/978-94-007-1451-9_1, © Springer Science+Business Media B.V. 2011

of the participants (whites and African Americans), the repository of blood specimens, much of which, unfortunately, perished in Hurricane Katrina, the numerous biochemical and physiological parameters, information diligently collected about features of growth and development, life styles and environmental factors, as well as genotypes and a host of cardiovascular phenotypes have provided a rich source of samples and data that enabled numerous investigations of the genesis of cardiovascular aging during the formative years and early adulthood. These same features provided the framework for studies of the mechanisms that fashion LTL dynamics in young adults. This chapter briefly reviews these studies, but before describing their findings and potential ramifications, it is essential to explain the biological underpinning of LTL dynamics.

1.2 LTL Dynamics Mirror Hematopoietic Stem Cell (HSC) Telomere Dynamics and Register the Inflammatory and Oxidative Stress Burdens

The basic question regarding the meaning of LTL dynamics has weighed on the discipline of telomere epidemiology from its very beginning. Like other human somatic cells, HSCs lack sufficient activity of telomerase [1–3], the enzyme that counteracts telomere shortening with each cell division [4]. Consequently, as HSC replicate their telomere length progressively shortens—a phenomenon that stems from the inability of DNA polymerase to replicate the lagging strand of DNA to its terminus [5]. Age-dependent telomere shortening in HSCs cannot be measured with current methodology, because HSCs are unavailable in sufficient quantities for routine measurements of their telomere lengths. Therefore, telomere shortening in granulocytes, which are post-mitotic cells with a short biological life, was originally used to model HSC replication kinetics and telomere dynamics [6]. However, recent research established that age-dependent shortening in the mean length of telomeres from all leukocytes, i.e., LTL, is as good a surrogate of HSC telomere dynamics as age-dependent telomere shortening in granulocytes [7, 8]. This was shown on two levels, first in newborns and then throughout the human life course.

Hematopoietic progenitor cells (HPCs) are much more abundant than HSCs and in newborns many of them circulate in the blood. Different leukocyte lineages in the newborn blood, including granulocytes, have similar telomere length as that of HPCs, and by inference telomere length of HSCs, which are situated slightly higher than HPCs in the hematopoietic system hierarchy [8]. Moreover, LTL is highly correlated with telomere length in granulocytes throughout the human lifespan [8]. It is safe to conclude, therefore, that due to the hierarchical nature of the hematopoietic system, LTL dynamics largely register telomere dynamics in HSCs [7].

LTL is highly variable among newborns [9–12] and its rate of shortening is rapid during the period of growth and development [7, 11–14]. That is evidently because the HSC pool expands in tandem with the growing soma and at the same time it must sustain the expansion of the HPC pool. These processes entail symmetric replications (each replicating HSC gives rise to two daughter HSCs) and asymmetric replications (each replicating HSC gives rise to one HSC and one HPC) [15], which result in rapid telomere shortening. During adulthood LTL shortening is much slower than earlier in life and it primarily reflects 'housekeeping' replicative activity of HSCs to accommodate the homeostatic needs of the individual.

How do LTL dynamics, reflecting HSC telomere dynamics, figure in the biology of aging and aging-related diseases? Clearly, HSC replication, expressed in age-dependent LTL shortening, charts the course of growth and development. But in addition, LTL shortening due to HSC replication evidently records the accruing burden of inflammation and oxidative stress. This unique ability stems from the fact that chronic inflammation engenders a greater expenditure of leukocytes, which must be accommodated for by increased HSC replication. Moreover, as telomeres are highly sensitive to the hydroxyl radical [16, 17], increased oxidative stress promotes more loss of telomeres repeats per each replication of HSCs. Thus, LTL attrition since birth is a record of not only growth and development but also the cumulative burden of inflammation and oxidative stress over the individual's life course. These features of LTL dynamics might explain the observed associations of shorten LTL with atherosclerosis [18–25], an aging-related disease that is marked by a chronic but indolent increase in inflammation and oxidative stress [26–28].

1.3 Insight Gained from Studying LTL Dynamics in the BHS

Most studies that explored the relations between LTL and various indices of aging were based on the cross-sectional model in which LTLs in persons of different ages were correlated with phenotypes of interest. But the longitudinal nature of the BHS has provided the opportunity to chart age-dependent LTL shortening in the individuals.

The BHS was the first to show the wide inter-individual variation in the rate of age-dependent LTL shortening [29, 30]. What's more, the study found that the rate of LTL attrition in the individual was correlated with the change in body mass index (BMI) over time, so that individuals with a greater weight gain had a faster rate of LTL shortening [29]. The link between BMI and other indices of obesity has been subsequently confirmed in other large-scale cross-sectional studies [31, 32]. The BMI-LTL connection might be mediated through insulin resistance [29, 33, 34], which is a state of heightened inflammation, or because the increase in the BMI itself. Inflammation might also explain the intriguing association of LTL with HDL-cholesterol that was observed in the Bogalusa Heart Study, as HDL-cholesterol [35] displays anti-inflammatory and anti-oxidant activities [36–38].

The BHS was also instrumental for the findings that LTL is longer in African Americans than whites [39], a finding replicated in the Family Heart Study and the Cardiovascular Health Study [39, 40]. This finding was unexpected, given that

LTL is shorter in whites who suffer from cardiovascular disease (CVD), and the greater susceptibility of African Americans to overall CVD. However, the LTL-CVD connection in whites is primarily due to a short LTL in patients with coronary atherosclerosis. Although African Americans are more susceptible than whites to hypertension [41, 42], left ventricular hypertrophy and kidney failure [43–48], they are less prone than whites to coronary atherosclerosis, as confirmed by coronary artery calcification studies [49–53]. Autopsy data from Bogalusa and that of PDAY show more vascular lesions in African Americans, but earlier and more coronary fibrous plaques in white males.

Although the underlying mechanisms for the longer LTL in African Americans are not known at present, they might relate to the benign ethnic neutropenia that is often displayed by African Americans [54–58]. Benign ethnic neutropenia evidently results from diminished recruitment of neutrophils from the bone marrow rather than increased neutrophil adherence in post-capillary venules [55, 57, 59]. This would entail diminished stimulus for HSC replication, particularly during growth and development, when HSC replication proceeds at a relatively rapid pace. Of note, benign ethnic neutropenia largely stems from variants of the Duffy Antigen Receptor for Chemokines (*DARC*) [60, 61] and a recent genome-wide association study, which included the Bogalusa Heart Study, showed that a locus that harbors the chemokine (C-X-C motif) receptor 4 gene (*CXCR4*) is associated with LTL in whites [62]. Both *DARC* and *CXCR4* encode proteins in control of neutrophil trafficking across the bone marrow.

The longitudinal evaluation of LTL dynamics in BHS participants had generated two intriguing observations that were subsequently replicated by others. First, the rate of LTL attrition was proportional to LTL at baseline [30, 35, 63], meaning that, everything else being equal, individuals with a longer LTL at baseline displayed a faster rate of age-dependent LTL shortening. Second a small subset of participants displayed LTL lengthening rather than LTL shortening during the follow-up period [29, 30, 33, 35]. The dependence of the rate of LTL shortening on LTL itself might be explained by the fact that oxidative stress accelerates telomere shortening. As longer telomeres are a bigger target for free radicals, the amount of telomere repeats that are clipped off with each replication might be greater for longer than shorter telomeres.

Although the original finding of LTL elongation observed in BHS participants was replicated by other longitudinal studies [64–68], a recent systematic evaluation of the underlying cause in BHS participants indicates that that LTL elongation is in fact an artifact that relates to the measurement error of LTL in relation to the duration of the follow-up [63]. This was shown by examining the relation between LTL elongation and the duration of follow-up and factoring the effect of the measurement error on this relation.

The notion that LTL lengthening with age is primarily the result of measurement error might be difficult to accept, since it was reported by different groups. But both theoretical considerations and the empirical data generated based on recent LTL data from BHS participants unequivocally indicate that this is the case under most circumstances. Elongation of LTL would suggest that HSC have switched to a telomere elongation mode and that LTL dynamics have somehow decoupled from age, both of which are highly unlikely.

1.4 Conclusions

By now, there is a vast and sometimes conflicting literature about the links of LTL with aging and aging-related diseases. That said, the original observations in the BHS were instrumental for the development of a whole new look at what LTL dynamics are all about and in what way they enlighten us about human aging.

References

1. Broccoli D, Young JW, de Lange T (1995) Telomerase activity in normal and malignant hematopoietic cells. Proc Natl Acad Sci U S A 92:9082–9086
2. Yui J, Chiu CP, Lansdorp PM (1998) Telomerase activity in candidate stem cells from fetal liver and adult bone marrow. Blood 91:3255–3262
3. Chiu CP, Dragowska W, Kim NW, Vaziri H, Yui J, Thomas TE, Harley CB, Lansdorp PM (1996) Differential expression of telomerase activity in hematopoietic progenitors from adult human bone marrow. Stem Cells 14:239–248
4. Blackburn EH (2005) Telomeres and telomerase: their mechanisms of action and the effects of altering their functions. FEBS Lett 579:859–862
5. Olovnikov AM (1996) Telomeres, telomerase, and aging: origin of the theory. Exp Gerontol 31:443–448
6. Shepherd BE, Guttorp P, Lansdorp PM, Abkowitz JL (2004) Estimating human hematopoietic stem cell kinetics using granulocyte telomere lengths. Exp Hematol 32:1040–1050
7. Sidorov I, Kimura M, Yashin A, Aviv A (2009) Leukocyte telomere dynamics and human hematopoietic stem cell kinetics during somatic growth. Exp Hematol 37:514–524
8. Kimura M, Gazitt Y, Cao X, Zhao X, Lansdorp PM, Aviv A (2010) Synchrony of telomere length among hematopoietic cells. Exp Hematol 38:854–859
9. Okuda K, Bardeguez A, Gardner JP, Rodriguez P, Ganesh V, Kimura M, Skurnick J, Awad G, Aviv A (2002) Telomere length in the newborn. Pediatr Res 52:377–381
10. Akkad A, Hastings R, Konje JC, Bell SC, Thurston H, Williams B (2006) Telomere length in small-for-gestational-age babies. Br J Obstet Gynaecol 113:318–323
11. Rufer N, Brümmendorf TH, Kolvraa S, Bischoff C, Christensen K, Wadsworth L, Schulzer M, Lansdorp PM (1999) Telomere fluorescence measurements in granulocytes and T lymphocyte subsets point to a high turnover of hematopoietic stem cells and memory T cells in early childhood. J Exp Med 190:157–167
12. Frenck RW Jr, Blackburn EH, Shannon KM (1999) The rate of telomere sequence loss in human leukocytes varies with age. Proc Natl Acad Sci U S A 95:5607–5610
13. Zeichner SL, Palumbo P, Feng Y, Xiao X, Gee D, Sleasman J, Goodenow M, Biggar R, Dimitrov D (1999) Rapid telomere shortening in children. Blood 93:2824–2830
14. Baerlocher GM, Rice K, Vulto I, Lansdorp PM (2007) Longitudinal data on telomere length in leukocytes from newborn baboons support a marked drop in stem cell turnover around 1 year of age. Aging Cell 6:121–123
15. Morrison SJ, Kimble J (2006) Asymmetric and symmetric stem-cell divisions in the development of cancer. Nature 441:1068–1074

16. Tchirkov A, Lansdorp PM (2003) Role of oxidative stress in telomere shortening in cultured fibroblasts from normal individuals and patients with ataxia-telangiectasia. Hum Mol Genet 12:227–232
17. Sitte N, Saretzki G, von Zglinicki T (1998) Accelerated telomere shortening in fibroblasts after extended periods of confluency. Free Radic Biol Med 24:885–893
18. Butt HZ, Atturu G, London NJ, Sayers RD, Bown MJ (2010) Telomere length dynamics in vascular disease: a review. Eur J Vasc Endovasc Surg 40:17–26 (21 May 2010)
19. Samani NJ, van der Harst P (2008) Biological ageing and cardiovascular disease. Heart 94:537–539
20. Oeseburg H, de Boer RA, van Gilst WH, van der Harst P (2010) Telomere biology in healthy aging and disease. Pflugers Arch 459:259–268
21. Brouilette SW, Moore JS, McMahon AD, Thompson JR, Ford I, Shepherd J, Packard CJ, Samani NJ (2007) Telomere length, risk of coronary heart disease, and statin treatment in the West of Scotland Primary Prevention Study: a nested case-control study. Lancet 369:107–114
22. van der Harst P, van der Steege G, de Boer RA, Voors AA, Hal AS, Mulder MJ, Van Gilst WH, van Veldhuissen DJ (2007) Telomere length of circulating leukocytes is decreased in patients with chronic heart failure. J Am Coll Cardiol 49:1459–1464
23. Benetos A, Gardner JP, Zureik M, Labat C, Xiaobin L, Adamopoulos C, Temmar M, Bean KE, Aviv A (2004) Short telomeres are associated with increased carotid artery atherosclerosis in hypertensive subjects. Hypertension 43:182–185
24. O'Donnell CJ, Demissie S, Kimura M, Levy D, Gardner JP, White C, D'Agostino RB, Wolf PA, Polak J, Cupples A, Aviv A (2008) Leukocyte telomere length and carotid artery intimal medial thickness: the Framingham Heart Study. Arterioscler Thromb Vasc Biol 28:1165–1171
25. Mainous AG 3rd, Codd V, Diaz VA, Schoepf UJ, Everett CJ, Player MS, Samani NJ (2010) Leukocyte telomere length and coronary artery calcification. Atherosclerosis 21:262–267
26. Weber C, Zernecke A, Libby P (2008) The multifaceted contributions of leukocyte subsets to atherosclerosis: lessons from mouse models. Nat Rev Immunol 8:802–815
27. Ross R (1999) Atherosclerosis—an inflammatory disease. N Engl J Med 340:115–126
28. Schleicher E, Friess U (2007) Oxidative stress, AGE, and atherosclerosis. Kidney Int Suppl 106:S17–S26
29. Gardner JP, Li S, Srinivasan SR, Chen W, Kimura M, Lu X, Berenson GS, Aviv A (2005) Rise in insulin resistance is associated with escalated telomere attrition. Circulation 111:2171–2177
30. Aviv A, Chen W, Gardner JP Kimura M, Brimacombe M, Cao X, Srinivasan SR, Berenson GS (2009) Leukocyte telomere dynamics: longitudinal findings among young adults in the Bogalusa Heart Study. Am J Epidemiol 169:323–329
31. Valdes AM, Andrew T, Gardner JP, Kimura M, Oelsner E, Cherkas LM, Aviv A, Spector TD (2005) Increased body mass and cigarette smoking are associated with short telomeres in women. Lancet 366:662–664
32. Prescott J, McGrath M, Lee IM, Buring JE, De Vivo I (2010) Telomere length and genetic analyses in population-based studies of endometrial cancer risk. Cancer 116:4275–4282
33. Demissie S, Levy D, Benjamin EJ, Cupples LA, Gardner JP, Herbert A, Kimura M, Larson MG, Meigs JB, Keaney JF, Aviv A (2006) Insulin resistance, oxidative stress, hypertension, and leukocyte telomere length in men from the Framingham Heart Study. Aging Cell 5:325–330
34. Atzmon G, Cho M, Cawthon RM, Budagov T, Katz M, Yang X, Siegel G, Bergman A, Huffman DM, Schechter CB, Wright WE, Shay JW, Barzilai N, Govindaraju DR, Suh Y (2010) Evolution in health and medicine Sackler colloquium: genetic variation in human telomerase is associated with telomere length in Ashkenazi centenarians. Proc Natl Acad Sci U S A 107(1):1710–1717
35. Chen W, Gardner JP, Kimura M, Brimacombe M, Cao X, Srinivasan SR, Berenson GS, Aviv A (2009) Leukocyte telomere length is associated with HDL cholesterol levels: the Bogalusa Heart Study. Atherosclerosis 205:620–625

36. Norata GD, Catapano AL (2005) Molecular mechanisms responsible for the anti-inflammatory and protective effect of HDL on the endothelium. Vasc Health Risk Manage 1:119–129
37. Negre-Salvayre A, Dousset N, Ferretti G et al (2006) Antioxidant and cytoprotective properties of high-density lipoproteins in vascular cells. Free Radic Biol Med 41:1031–1040
38. Ansell BJ (2007) Targeting the anti-inflammatory effects of high-density lipoprotein. Am J Cardiol 100(11 A):3–9
39. Hunt SC, Chen W, Gardner JP, Kimura M, Srinivasan SR, Eckfeldt JH, Berenson GS, Aviv A (2008) Leukocyte telomeres are longer in African Americans than in whites: the National Heart, Lung, and Blood Institute Family Heart Study and the Bogalusa Heart Study. Aging Cell 7:451–458
40. Fitzpatrick AL, Kronmal RA, Kimura M, Gardne JP, Psaty BM, Jenny NS, Tracy RP, Hardikar S, Aviv A Leukocyte telomere length and mortality in the cardiovascular health study. J Gerontol Biol Sci Med Sci (in press)
41. Grim CE, Robinson M (1996) Blood pressure variation in blacks: genetic factors. Semin Nephrol 16:83–93
42. Nesbitt SD (2005) Hypertension in black patients: special issues and considerations. Curr Hypertens Rep 7:244–248
43. Stewart AD, Millasseau SC, Dawes M, Kyd PA, Chambers JB, Ritter JM, Chowienczyk PJ (2006) Aldosterone and left ventricular hypertrophy in Afro-Caribbean subjects with low renin hypertension. Am J Hypertens 9:19–24
44. El-Gharbawy AH, Nadig VS, Kotchen JM, Grim CE, Sagar KB, Kaldunski M, Hamet P, Pausova Z, Gaudet D, Gossard F, Kotchen TA (2001) Arterial pressure, left ventricular mass, and aldosterone in essential hypertension. Hypertension 37:845–850
45. Moe GW, Tu J (2010) Heart failure in the ethnic minorities. Curr Opin Cardiol 25:124–130
46. Yancy CW, Strong M (2004) The natural history, epidemiology, and prognosis of heart failure in African Americans. Congest Heart Fail 10:15–18
47. Martins D, Tareen N, Norris KC (2002) The epidemiology of end-stage renal disease among African Americans. Am J Med Sci 323:65–71
48. Martínez-Maldonado M (2001) Role of hypertension in the progression of chronic renal disease. Nephrol Dial Transplant 16(1):63–66
49. Tang W, Detrano RC, Brezden OS, Georgiou D, French WJ, Wong ND, Doherty TM, Brundage BH (1995) Racial differences in coronary calcium prevalence among high-risk adults. Am J Cardiol 75:1088–1091
50. Aiyer AN, Kip KE, Marroquin OC, Mulukutla SR, Edmundowicz D, Reis SE (2007) Racial differences in coronary artery calcification are not attributed to differences in lipoprotein particle sizes: the heart strategies concentrating on risk evaluation (Heart SCORE) Study. Am Heart J 153:328–324
51. Detrano R, Guerci AD, Carr JJ et al (2008) Coronary calcium as a predictor of coronary events in four racial ethnic groups. New Engl J Med 358:1336–1345
52. LaMonte MJ, FitzGerald SJ, Church TS et al (2005) Coronary artery calcium score and coronary heart disease events in a large cohort of asymptomatic men and women. Am J Epidemiol 162:421–429
53. Arad Y, Goodman KJ, Roth M, Newstein D, Guerci AD (2005) Coronary calcification, coronary disease risk factors, C-reactive protein, and atherosclerotic cardiovascular disease events: the St. Francis Heart Study. J Am Coll Cardiol 46:158–165
54. Mayr FB, Spiel AO, Leitner JM, Firbas C, Kliegel T, Jilma B (2007) Ethnic differences in plasma levels of interleukin-8 (IL-8) and granulocyte colony stimulating factor (G-CSF). Trans Res 149:10–14
55. Phillips D, Rezvani K, Bain BJ (2000) Exercise induced mobilization of marginated granulocyte pool in the investigation of ethnic neutropenia. J Clin Pathol 53:481–483
56. Haddy TB, Rana SR, Castro O (1999) Benign ethnic neutropenia: what is a normal absolute neutrophil count? J Lab Clin Med 133:15–22

57. Bain BJ, Phillips D, Thomson K, Richardson D, Gabriel I (2000) Investigation of the effect of marathon running on leukocyte counts of subjects of different ethnic origins: relevance to aetiology of ethnic neutropenia. Br J Haemotol 108:483–487
58. Broun GO Jr, Herbig FK, Hamilton JR (1966) Leukopenia in Negroes. N Engl J Med 275:1410–1413
59. Athens JW, Raab SO, Haaab OP, Mauer AM, Ashenbrucker H, Cartwright GE, Wintrobe MM (1961) Leukokinetic studies III. The distribution of granulocytes in the blood of normal subjects. J Clin Invest 40:159–164
60. Reich D, Nalls MA, Kao WH, Akylbekova EL, Tandon A, Patterson N, Mullikin J, Hsueh WC, Cheng CY, Coresh J, Boerwinkle E, Li M, Waliszewska A, Neubauer J, Li R, Leak TS, Ekunwe L, Files JC, Hardy CL, Zmuda JM, Taylor HA, Ziv E, Harris TB, Wilson JG (2009) Reduced neutrophil count in people of African descent is due to a regulatory variant in the Duffy antigen receptor for chemokines gene. PLoS Genet 5:e1000360
61. Nalls MA, Wilson JG, Patterson NJ, Tandon A, Zmuda JM, Huntsman S, Garcia M, Hu D, Li R, Beamer BA, Patel KV, Akylbekova EL, Files JC, Hardy CL, Buxbaum SG, Taylor HA, Reich D, Harris TB, Ziv E (2008) Admixture mapping of white cell count: genetic locus responsible for lower white blood cell count in the Health ABC and Jackson Heart studies. Am J Hum Genet 82:81–87 (Erratum in: Am J Hum Genet 2008; 82: 532)
62. Levy D, Neuhausen BL, Hunt SC, Kimura M, Hwang S-H, Chen W, Bis JC, Fitzpatrick AL, Smith E, Andrew D, Gardner JP, Srinivasan SR, Schork N, Rotter JI, Herbig U, Psaty, BM, Sastrasinh M, Murray SS, Vasan RS, Province MA, Glazer NL, Lu X, Cao X, Kronmal R, Mangino M, Soranzo N, Spector TD, Berenson GS, Aviv A (2010) Genome-wide association identifies OBFC1 as a locus involved in human leukocyte telomere biology. Proc Nat Acad Sci U S A 107:9293–9298
63. Chen W, Kimura M, Kim S, Cao X, Srinivasan SR, Berenson GS, Kark JD, Aviv A. Longitudinal vs. cross-sectional evaluations of leukocyte telomere length dynamics: age-dependent telomere shortening is the rule. J Gerontol Biol Sci Med Sci (in press)
64. Martin-Ruiz CM, Gussekloo J, van Heemst D et al (2005) Telomere length in white blood cells is not associated with morbidity or mortality in the oldest old: a population-based study. Aging Cell 4:287–290
65. Ehrlenbach S, Willeit P, Kiechl S et al (2009) Influences on the reduction of relative telomere length over 10 years in the population-based Bruneck Study: introduction of a well-controlled high-throughput assay. Int J Epidemiol 38:1725–1734
66. Nordfjäll K, Svenson U, Norrback KF, Adolfsson R, Lenner P, Roos G (2009) The individual blood cell telomere attrition rate is telomere length dependent. PLoS Genet 5:e1000375
67. Farzaneh-Far R, Lin J, Epel E, Lapham K, Blackburn E, Whooley MA (2010) Telomere length trajectory and its determinants in persons with coronary artery disease: longitudinal findings from the heart and soul study. PloS One 5:e8612
68. Farzaneh-Far R, Lin J, Epel ES, Harris WS, Blackburn EH, Whooley MA (2010) Association of marine omega-3 fatty acid levels with telomeric aging in patients with coronary heart disease. JAMA 303:250–257

Chapter 2
Fetal Origins of Variables Related to Cardio-Metabolic Risk

Sathanur R. Srinivasan

Abstract Low birth weight for gestational age, regardless of socio-economic background and geographical location, is considered a risk factor for chronic diseases such as cardiovascular disease and type 2 diabetes mellitus in later life. This overview focuses on the racial (black–white) divergence in birth weight and adverse effects of low birth weight on aspects of cardio-metabolic risk involving anthropometric, hemodynamic, metabolic and inflammatory variables during growth periods of childhood, adolescence, and adulthood along with pulsatile behavior of the vasculature in adulthood noted in the Bogalusa Heart Study cohort. Several putative mechanisms linking these adverse relationships are discussed, thereby providing a rationale for primordial prevention.

Keywords Arterial stiffness • Cardio-metabolic risk • CV risk factor • Fetal growth • Inflammatory marker • Low birth weight • Racial difference

2.1 Introduction

Low birth weight for gestational age is considered to be an indicator of metabolic imprinting and developmental plasticity associated with a compromised intrauterine growth and development [1–5]. Developmental plasticity reflects gene expression mediated in part by epigenetic processes in response to environmental factors and subsequent risk of diseases [6]. Studies world-wide, regardless of socio-economic background, have linked low birth weight to increased risk of developing insulin resistance, dyslipidemia, hypertension, coronary heart disease, and type 2 diabetes [7–10], although some investigators question this relationship [11, 12]. This chapter highlights the effect of low birth weight on aspects of cardio-metabolic risk noted in the Bogalusa Heart Study biracial (black–white) cohort [13–22].

S. R. Srinivasan (✉)
Department of Epidemiology, Biochemistry, Center for Cardiovascular Health, Tulane University School of Public Health and Tropical Medicine, New Orleans, LA, USA
e-mail: ssriniv1@tulane.edu

G. S. Berenson (ed.), *Evolution of Cardio-Metabolic Risk from Birth to Middle Age*,
DOI 10.1007/978-94-007-1451-9_2, © Springer Science+Business Media B.V. 2011

2.2 Racial (Black–White) Difference

Birth weight of children born to Bogalusa residents between January, 1974 and June, 1975 were examined by race. Of these children, 100% blacks and 97.5% whites participated in the study [13]. As illustrated in Fig. 2.1, white children at birth when compared to black children consistently weighed more. Ninety percent of white newborns weighed between 2353 g (5th percentile) and 4186 g (95th percentile); whereas for black newborns the corresponding values were 1962 g and 3989 g, respectively. This black–white contrast in birth weight for gestational age was also seen in later studies, including our own [15, 18, 23, 24].

The effect that socio-economic and genetic factors may have on the birth weight of these two racial groups were indirectly assessed by comparing the birth weight data by race of children born in Bogalusa public vs private hospitals. The racial make-up of children born at the state hospital was 66% black and 34% white; at the private hospital, 19% black and 81% white. At both hospitals, white neonates weighed significantly more than their black counterparts, although in both races those born at the private hospital weighed significantly more than those born at the public hospital. Further, mean birth weight of blacks born at the private hospital was almost identical to the mean birth weight of whites born at the public hospital. Taken together, these findings indicate that both socio-economic and genetic factors influence the weight at birth.

2.3 Relation to Anthropometric, Metabolic, and Hemodynamic Variables

Earlier studies have linked low birth weight to adverse levels of cardiovascular risk factor variables in childhood and adolescence [25–29]. However, information is scant on data linking low birth weight to longitudinal trends of adiposity, blood pressure, lipids, and glucose homeostasis variables (glucose, insulin, and insulin

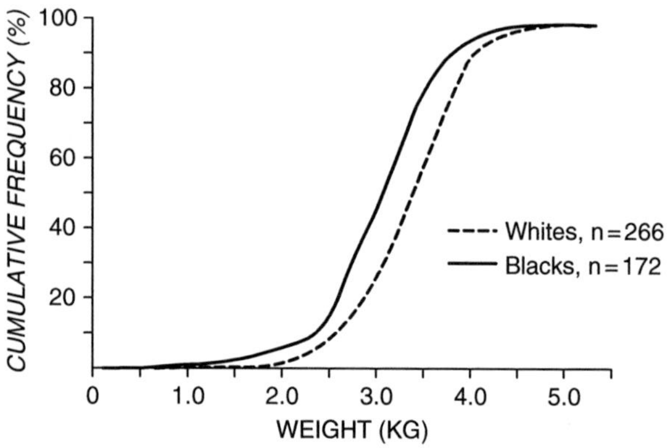

Fig. 2.1 Cumulative frequency distributions for weight at birth of black and white Bogalusa newborns: the Bogalusa Heart Study. [13]

Table 2.1 Levels (mean ± SD) of risk variables during childhood and adolescence by birth weight: the Bogalusa Heart Study. [14]

Variable	Childhood (4–11 years)		Adolescence (12–18 years)	
	Low birth weight	Control	Low birth weight	Control
BMI (kg/m^2)	16.7±2.6	17.5±2.9	21.6±5.2	22.7±5.0
Subsc. skinfold (mm)	8.2±5.1	8.3±6.4	15.9±10.0	16.0±10.7
Syst. BP (mm Hg)	96.4±9.0	98.6±8.1	108.8±9.1	105.2±8.7
Diast. BP (mm Hg)	57.2±10.2	58.7±7.9	66.1±8.0	66.4±7.4
Triglycerides (mg/dL)	62.6±23.2	52.8±20.3	87.1±29.0	83.2±35.3
HDL cholesterol (mg/dL)	44.3±22.6*	54.7±17.4	49.9±12.8	51.2±11.5
LDL cholesterol (mg/dL)	76.0±35.4*	68.6±37.6	99.5±24.7	98.4±24.3
Glucose (mg/dL)	79.7±8.1	80.9±9.7	85.4±8.2**	81.6±7.4
Insulin (μU/mL)	8.5±5.7	7.4±4.6	14.8±7.5	13.2±8.6
HOMA-IR	1.7±1.2	1.6±1.0	3.0±2.4	2.7±2.0

Difference between groups (adjusted for age, race, and gender), *p = 0.05; **p = 0.02
HOMA-IR homeostasis model assessment index of insulin resistance

resistance index) measured simultaneously and serially during the developmental periods of childhood and adolescence. The Bogalusa Heart Study subjects followed from childhood to adolescence by repeated surveys were categorized into singleton and full term low birth weight (below the gestation age- and race-specific 10th percentile) and control (between the gestation age- and race-specific 50th and 75th percentiles) groups [14]. As shown in Table 2.1, low birth weight vs control group had significantly lower mean high-density lipoprotein (HDL) cholesterol and higher low-density lipoprotein (LDL) cholesterol during childhood (ages 4–11 years); higher glucose during adolescence (ages 12–18 years). In addition, yearly rates of change from childhood to adolescence in systolic blood pressure, LDL cholesterol, and glucose were faster, and body mass index (BMI) slower among the low birth weight group. In a multivariate analysis of the serial data, presented in Table 2.2, the

Table 2.2 Independent association of low birth weight with longitudinal trends of systolic blood pressure, triglycerides and glucose from childhood to adolescence. [14]

Independent variables retained	Syst. BP		Triglycerides		Glucose	
	β†	p-value	β	p-value	β	p-value
Birth weight (low vs control)	3.84	0.02	48.6	0.08	15.20	0.07
Gender (male vs female)	–	–	–	–	4.31	<0.001
Age	0.39	0.40	12.21	0.01	6.60	<0.001
Age^2	0.06	0.002	−0.44	0.04	−0.31	<0.001
Insulin	0.11	0.03	1.16	<0.001	0.22	0.02
BMI	0.34	<0.001	–	–	–	–
Birth weight × age	0.45	0.004	12.71	0.03	0.10	0.07
Birth weight × age^2	–	–	0.55	0.02	2.65	0.10

† GEE regression coefficient. The model included birth weight (low vs control) along with age, age^2, race, and gender, and their interaction with birth weight; and risk variable as applicable

independent adverse effects of low birth weight on the longitudinal trends of systolic blood pressure, triglycerides, and glucose were discernable in the study cohort, regardless of race and sex. An analysis of the data set using the World Health Organization criteria for low birth weight (<2500 g) gave essentially the same results. Of note, observed adverse trends associated with low birth weight group vs control group were influenced by age in that the difference became greater in magnitude as the children got older. Earlier studies have found such strong associations with increasing age [30]. Further, although the rate of yearly increase in BMI, which also includes muscle mass, was significantly lower in low birth weight group, the rate of increase in subscapular skinfold, a measure of truncal fat, remained similar to that of the control group. This suggests a gaining of truncal fat, in relative term, over muscle mass in the low birth weight group.

In terms of hemodynamic variables, additional studies on the developmental trends from childhood to mid-adulthood indicated that birth weight was independently inversely associated with systolic blood pressure as well as diastolic blood pressure and pulse pressure [17]. The interaction of birth weight with BMI was negative for all these three hemodynamic variables, but the trend was significant only for systolic blood pressure. Figure 2.2 illustrates the relation between birth weight and later systolic blood pressure (at age≥ 35 years) within each tertile of adult BMI. That adjustment for later BMI strengthens this association indicates BMI as a negative confounder as originally proposed by Gillman [31]. Importantly, the magnitude of the birth weight–systolic blood pressure relationship was significantly amplified with increasing age from pre-adolescence to mid-adulthood, regardless of adjustment for current BMI and race [18]. These finding are consistent with previous reports [30, 32]. Further, in this cohort, low birth weight was significantly associated with increased within-individual blood pressure variability assessed in terms of standard deviations of six serial blood pressure measurements from childhood to adulthood. In a multivariate model adjusted for race, sex and average age and blood pressure from childhood to adulthood, 1 kg of lower birth weight was associated with

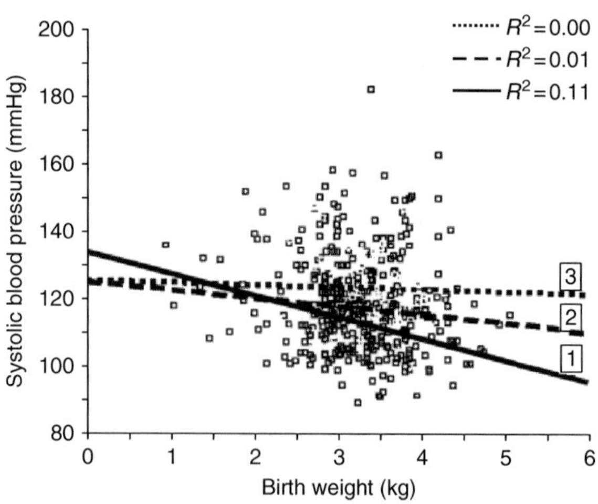

Fig. 2.2 The interaction of birth weight with body mass index (BMI) on systolic blood pressure at age≥35 years. BMI: *1* lowest tertile, *2* middle tertile, *3* highest tertile: the Bogalusa Heart Study. [15]

0.43 mm Hg increase in standard deviation of systolic blood pressure ($p=0.023$); 0.55 mm Hg of diastolic blood pressure ($p=0.002$). These observations suggest that the fetal programming and the increasing burden with age of unhealthy lifestyle behaviors affect the development of adult hypertension in a synergistic manner.

It is well established that genetics play a major role in determining both blood pressure levels and birth weight [33, 34]. Candidate genes and chromosomal regions have been identified for birth weight [35–37]. The genetic influence of β-adrenergic receptor gene polymorphisms (β2-AR Arg 16 Gly and β3-AR Trp 64 Arg), important modulators of fetal growth, sympathetic nervous system, hemodynamic status, obesity, and insulin sensitivity [38–41], on the inverse relation of birth weight to longitudinal changes in blood pressure has been found in the Bogalusa cohort, aged 21–47 years [19]. Blacks showed significantly lower birth weight and frequencies of β2-AR Gly 16 and β3-AR Trp 64 alleles and higher blood pressure levels and age-related trends than whites. Importantly, the strength of the birth weight—blood pressure association, measured as regression coefficients, was significantly modulated by the combination of β2-AR and β3-AR genotypes for systolic blood pressure ($p=0.042$ for interaction) and diastolic blood pressure ($p=0.039$ for interaction) age-related trend, with blacks and whites showing a similar trend in the interaction. Further, low birth weight was adversely related to longitudinal changes in BMI from childhood to adulthood depending on fat mass and obesity risk associated (FTO) genotype in the Bogalusa cohort [42]. These findings underscore the role of genetic factors in the fetal origins of hypertension later in life.

2.4 Relation to Inflammation

It is well recognized that cardio-metabolic syndrome and related CV diseases and type 2 diabetes all share inflammation as a common pathologic trait [43–45]. Further, a few studies have linked low birth weight for gestational age to increases in systemic inflammation as depicted by C-reactive protein [46, 47]. The influence of low birth weight on the white blood cell (WBC) count, a widely used biomarker of systemic inflammation [48, 49], during different growth periods of childhood (ages 4–11 years), adolescence (ages 12–17 years), and adulthood (ages 18–38 years) was examined in the Bogalusa cohort [20]. As illustrated in Fig. 2.3, the WBC count significantly decreased with increasing quartiles of birth weight in children and adults of both races. This inverse relationship was independent of age, race, sex, BMI, and smoking status. Adolescents, regardless of race, showed no such significant trend in this relationship. Because the pubertal period of growth and maturation is characterized by marked changes in fat mass and distribution along with insulin sensitivity [50–52], known strong correlates of inflammation [44, 48], it is likely that low birth weight may not be a strong enough proxy for prenatal factors influencing the WBC count during adolescence. These observations by showing inverse association between birth weight WBC count provide further rationale for linking fetal growth retardation to CV diseases and type 2 diabetes, and implicating an impact of inflammation.

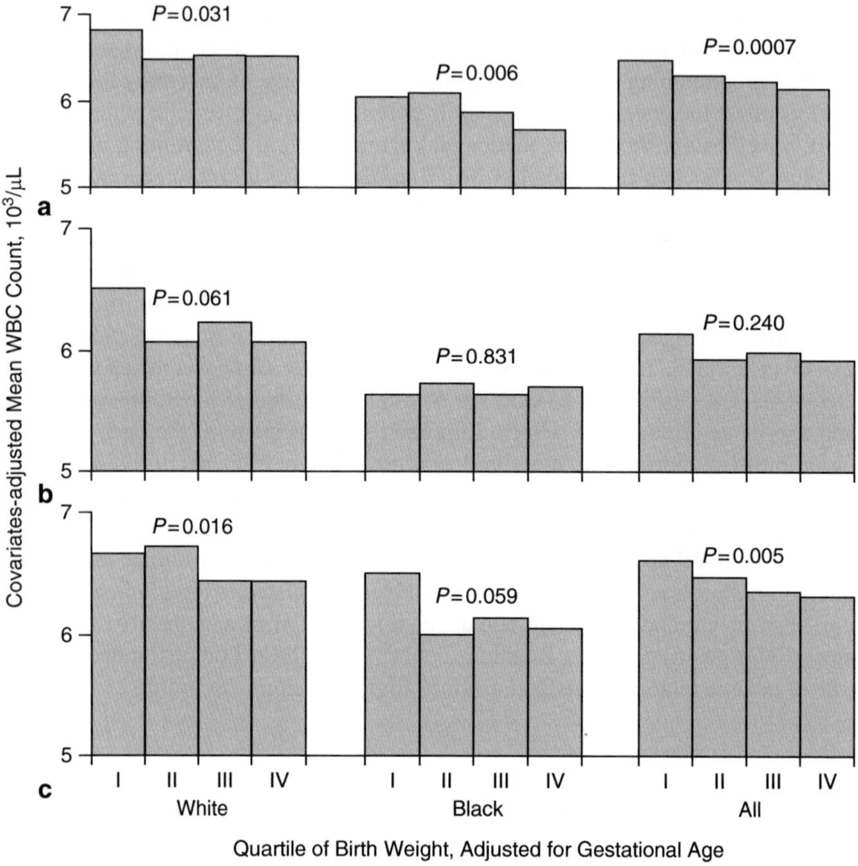

Fig. 2.3 Covariate-adjusted mean values of white blood cell (WBC) count by race- and sex-specific quartiles of gestational age-adjusted birth weight. Quartiles *I* and *IV* represent the lowest and highest birth weights, respectively, in children (**a**), adolescents (**b**), and adults (**c**). Covariates included age, sex, race (total sample), body mass index, and smoking status (adolescents and adults). The Bogalusa Heart Study. [20]

2.5 Relation to Arterial Stiffness

Recent studies have shown that alterations in the pulsatile behavior of the vasculature may be a sensitive marker to detect vascular dysfunction related to CV risk factors and mortality [53–55]. Therefore, studies of arterial compliance in terms of pulsatile behavior can help determine the role of birth weight in relation to CV risk. This aspect was examined in the Bogalusa cohort of asymptomatic younger adults aged 18–44 years [21]. Subjects were categorized into low (below the race–sex specific 10th percentile) and normal (between the race–sex specific 10th and 90th

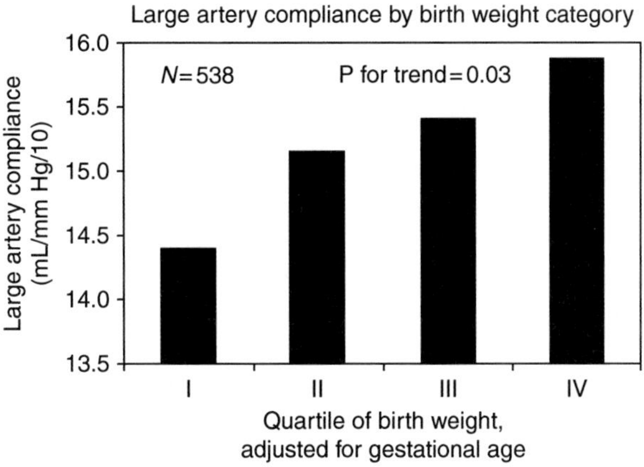

Fig. 2.4 Covariates-adjusted mean values of large artery compliance by race–sex specific gestational age-adjusted birth weight. Quartile *I* represents lowest birth weight; *IV* highest birth weight. Covariates included race, sex, and age. The observations show the relation of low birth weight and loss of vascular compliance and early vascular stiffness. The Bogalusa Heart Study. [21]

percentiles) birth weight groups. Pulsatile arterial function was assessed in terms of large artery compliance and small artery compliance. A modified Windkessel model of the circulation was used to quantify changes in arterial waveform morphology in terms of large artery (capacitive) compliance, representative of the aorta and major branches, and small artery (oscillatory) compliance, representative of the distal part of the circulation including the arteriolar bed, and systemic vascular resistance [56]. Low vs normal birth weight group had significantly lower large artery compliance. Further, in the total sample, after adjusting for gestational age, race and sex, the large artery compliance increased significantly across quartiles of increasing birth weight specific for race, sex, and gestational age (Fig. 2.4). This relationship was independent of race, sex, and adulthood age, BMI, body surface area, systolic blood pressure, diastolic blood pressure, pulse rate, triglycerides/HDL cholesterol ratio, insulin resistance index and smoking status. No such adverse independent effect of low birth weight was noted with respect to small artery compliance and systemic vascular resistance.

The observed positive association between birth weight and large artery compliance is consistent with earlier studies [57–60], except Atherosclerosis Risk in Young Adults study in somewhat older subjects [61]. The lack of association between birth weight and small artery compliance may be due to the fact that the association between birth weight and compliance was found to be stronger in elastic arteries than the small muscular arteries [57, 62]. Further, birth weight was also inversely and independently related to both systolic blood pressure and brachial-ankle pulse wave velocity, another marker of arterial stiffness related to elastic properties of aorta and its larger branches, measured at younger to mid-adulthood in the Bogalusa cohort [22]. Because arterial stiffness is both a cause and result of hypertension [63, 64], increased arterial stiffness may be a link between the inverse association between birth weight and blood pressure later in life, also to developing systolic hypertension and increasing pulse pressure. These findings, regardless of underlying mechanisms, highlight the adverse CV risk associated with low birth weight.

2.6 Mechanisms

Several putative mechanisms link low birth weight to the adverse cardio-metabolic risk profile [65]. It has been suggested that insulin resistance may be one mechanism by which intrauterine events may program disease risk [66]. Under-nutrition in utero is known to cause permanent impairment in growth, structure, and function of, among others, muscle [66, 67], fat [68], endocrine pancreas [69, 70], liver [71], renal nephrons [72, 73], sympathetic nervous system [74], glucocorticoid hormones [75], hypothalamic–pituitary–adrenal axis [76] and vasculature [77] due to adaptive programming, resulting in adverse profiles of cardio-metabolic risk factor variables and related disorders later in life [7, 9, 65]. Of interest, since there is little muscle cell replication after birth [78], low birth weight individuals under a nutritionally rich environment later in life will develop a disproportionately high fat mass and related state of chronic low grade inflammation induced by adipose tissue cells including monocytes [65, 79]. With respect to large artery compliance, it has been proposed that impairment of synthesis of elastin and changes in collagen along with endothelial dysfunction in large arteries during the period of intrauterine growth retardation in low birth weight babies result in permanent changes in mechanical properties of these vessels [80, 81]. In addition, structural adaptations in the large artery wall occur in response to alterations in Doppler blood flow velocity waveforms under conditions of intrauterine growth retardations [61].

2.7 Conclusion

Low birth weight, albeit a crude surrogate marker for fetal growth and maturation, is a potential early risk factor for the emergence of anthropometric, metabolic and hemodynamic disorders and related diseases [2, 65]. In terms of public health, the importance of primordial prevention is obvious. As stated by Barker and colleagues [2, 7], primary prevention lies in protecting fetal development. This entails providing comprehensive prenatal care services, taking into account differences in the prevalence of low birth weight babies among racial groups in the general population.

References

1. Barker DJ (1994) Programming the baby. In: Barker DJ (ed) Mothers, babies and disease later in life. BMJ Publishing Group, London, 14–36
2. Barker DJ (2001) Fetal origins of cardiovascular and lung disease. Marcel Dekker, New York, pp 1–197
3. Waterland RA, Garza C (1999) Potential mechanisms of metabolic imprinting that lead to chronic disease. Am J Clin Nutr 69:179–197
4. Hattersley AT, Tooke JE (1999) The fetal insulin hypothesis: an alternative explanation of the association of low birthweight with diabetes and vascular disease. Lancet 353:1789–1792

5. Barker DJP (2004) Developmental origins of adult health and disease. J Epidemiol Commun Health 58:114–115
6. Gluckman PD, Hanson MA, Cooper C, Thornburg KL (2008) Effect of in utero and early-life conditions on adult health and disease. N Engl J Med 359:61–73
7. Barker DJ, Hales CN, Fall CH, Osmond C, Phipps K, Clark PM (1993) Type 2 (non-insulin-dependent) diabetes mellitus, hypertension and hyperlipidaemia (syndrome X): relation to reduced fetal growth. Diabetologia 36:62–67
8. Valdez R, Athens MA, Thompson GH, Bradshaw BS, Stern MP (1994) Birthweight and adult health outcomes in a biethnic population in the USA. Diabetologia 37:624–631
9. Leon DA, Lithell HO, Vâgerö D, Koupilová I, Mohsen R, Berglund L, Lithell UB, McKeigue PM (1998) Reduced fetal growth rate and increased risk of death from ischaemic heart disease: cohort study of 15,000 Swedish men and women born 1915–29. Br Med J 317:241–245
10. Godfrey KM, Barker DJ (2000) Fetal nutrition and adult disease. Am J Clin Nutr 71(Suppl 5):1344S–1352S
11. Susser M, Levin B (1999) Ordeals for the fetal programming hypothesis. The hypothesis largely survives one ordeal but not another. Br Med J 318:885–886
12. Huxley R, Neil A, Collins R (2002) Unravelling the fetal origins hypothesis: is there really an inverse association between birthweight and subsequent blood pressure? Lancet 360:659–665
13. Frerichs RR, Srinivasan SR, Webber LS, Rieth MC, Berenson GS (1978) Serum lipids an lipoproteins at birth in a biracial population: the Bogalusa Heart Study. Pediatr Res 12:858–863
14. Frontini MG, Srinivasan SR, Xu J, Berenson GS (2004) Low birth weight and longitudinal trends of cardiovascular risk factor variables from childhood to adolescence: the Bogalusa Heart Study. BMC Pediatr 4:22
15. Mzayek F, Sherwin R, Fonseca V, Valdez R, Srinivasan SR, Cruickshank JK, Berenson GS (2004) Differential association of birth weight with cardiovascular risk variables in African–Americans and Whites: the Bogalusa Heart Study. Ann Epidemiol 14:258–264
16. Cruickshank JK, Mzayek F, Liu L, Kieltyka L, Sherwin R, Webber LS, Srinavasan SR, Berenson GS (2005) Origins of the "black/white" difference in blood pressure: roles of birth weight, postnatal growth, early blood pressure, and adolescent body size: the Bogalusa Heart Study. Circulation 111:1932–1937
17. Mzayek F, Hassig S, Sherwin R, Hughes J, Chen W, Srinivasan S, Berenson G (2007) The association of birth weight with developmental trends in blood pressure from childhood through mid-adulthood: the Bogalusa Heart study. Am J Epidemiol 166:413–420
18. Chen W, Srinivasan SR, Berenson GS (2010) Amplification of the association between birthweight and blood pressure with age: the Bogalusa Heart Study. J Hypertens 28:2046–2052
19. Chen W, Srinivasan SR, Hallman DM, Berenson GS (2010) The relationship between birthweight and longitudinal changes of blood pressure is modulated by beta-adrenergic receptor genes: the Bogalusa Heart Study. J Biomed Biotechnol 2010:543514
20. Chen W, Srinivasan SR, Berenson GS (2009) Influence of birth weight on white blood cell count in biracial (black–white) children, adolescents, and young adults: the Bogalusa Heart Study. Am J Epidemiol 169:214–218
21. Bhuiyan AR, Chen W, Srinivasan SR, Azevedo MJ, Berenson GS (2010) Relationship of low birth weight to pulsatile arterial function in asymptomatic younger adults: the Bogalusa Heart Study. Am J Hypertens 23:168–173
22. Mzayek F, Sherwin R, Hughes J, Hassig S, Srinivasan S, Chen W, Berenson GS (2009) The association of birth weight with arterial stiffness at mid-adulthood: the Bogalusa Heart Study. J Epidemiol Commun Health 63:729–733
23. Alexander GR, Kogan MD, Himes JH (1999) 1994–1996 U.S. singleton birth weight percentiles for gestational age by race, Hispanic origin, and gender. Matern Child Health J 3:225–231
24. Shiono PH, Klebanoff MA, Graubard BI, Berendes HW, Rhoads GG (1986) Birth weight among women of different ethnic groups. J Am Med Assoc 255:48–52

25. Uiterwaal CS, Anthony S, Launer LJ, Witteman JC, Trouwborst AM, Hofman A, Grobbee DE (1997) Birth weight, growth, and blood pressure: an annual follow-up study of children aged 5 through 21 years. Hypertension 30:267–271
26. Whincup PH, Cook DG, Shaper AG (1989) Early influences on blood pressure: a study of children aged 5–7 years. Br Med J 299:587–591
27. Bavdekar A, Yajnik CS, Fall CH, Bapat S, Pandit AN, Deshpande V, Bhave S, Kellingray SD, Joglekar C (1999) Insulin resistance syndrome in 8-year-old Indian children: small at birth, big at 8 years, or both? Diabetes 48:2422–2429
28. Li C, Johnson MS, Goran MI (2001) Effects of low birth weight on insulin resistance syndrome in Caucasian and African–American children. Diabetes Care 24:2035–2042
29. Murtaugh MA, Jacobs DR Jr, Moran A, Steinberger J, Sinaiko AR (2003) Relation of birth weight to fasting insulin, insulin resistance, and body size in adolescence. Diabetes Care 26:187–192
30. Law CM, de Swiet M, Osmond C, Fayers PM, Barker DJ, Cruddas AM, Fall CH (1993) Initiation of hypertension in utero and its amplification throughout life. Br Med J 306:24–27
31. Gillman MW (2002) Epidemiological challenges in studying the fetal origins of adult chronic disease. Int J Epidemiol 31:294–299
32. Gamborg M, Byberg L, Rasmussen F, Andersen PK, Baker JL, Bengtsson C, Canoy D, Drøyvold W, Eriksson JG, Forsén T, Gunnarsdottir I, Järvelin MR, Koupil I, Lapidus L, Nilsen TI, Olsen SF, Schack-Nielsen L, Thorsdottir I, Tuomainen TP, Sørensen TI, NordNet Study Group (2007) Birth weight and systolic blood pressure in adolescence and adulthood: meta-regression analysis of sex- and age-specific results from 20 Nordic studies. Am J Epidemiol 166:634–645
33. La Batide-Alanore A, Trégouët DA, Jaquet D, Bouyer J, Tiret L (2002) Familial aggregation of fetal growth restriction in a French cohort of 7,822 term births between 1971 and 1985. Am J Epidemiol 156:180–187
34. Luft FC (2000) Molecular genetics of human hypertension. Curr Opin Nephrol Hypertens 9:259–266
35. Pihlajamäki J, Vanhala M, Vanhala P, Laakso M (2004) The Pro12Ala polymorphism of the PPAR gamma 2 gene regulates weight from birth to adulthood. Obes Res 12:187–190
36. Arya R, Demerath E, Jenkinson CP, Göring HH, Puppala S, Farook V, Fowler S, Schneider J, Granato R, Resendez RG, Dyer TD, Cole SA, Almasy L, Comuzzie AG, Siervogel RM, Bradshaw B, DeFronzo RA, MacCluer J, Stern MP, Towne B, Blangero J, Duggirala R (2006) A quantitative trait locus (QTL) on chromosome 6q influences birth weight in two independent family studies. Hum Mol Genet 15:1569–1579
37. Johnston LB, Clark AJ, Savage MO (2002) Genetic factors contributing to birth weight. Arch Dis Child Fetal Neonatal Ed 86:F2–F3
38. Zhu H, Poole J, Lu Y, Harshfield GA, Treiber FA, Snieder H, Dong Y (2005) Sympathetic nervous system, genes and human essential hypertension. Curr Neurovasc Res 2:303–317
39. Wang X, Cui Y, Tong X, Ye H, Li S (2004) Effects of the Trp64Arg polymorphism in the beta3-adrenergic receptor gene on insulin sensitivity in small for gestational age neonates. J Clin Endocrinol Metab 89:4981–4985
40. Jaquet D, Trégouët DA, Godefroy T, Nicaud V, Chevenne D, Tiret L, Czernichow P, Lévy-Marchal C (2002) Combined effects of genetic and environmental factors on insulin resistance associated with reduced fetal growth. Diabetes 51:3473–3478
41. Ellsworth DL, Coady SA, Chen W, Srinivasan SR, Elkasabany A, Gustat J, Boerwinkle E, Berenson GS (2002) Influence of the beta2-adrenergic receptor Arg16Gly polymorphism on longitudinal changes in obesity from childhood through young adulthood in a biracial cohort: the Bogalusa Heart Study. Int J Obes Relat Metab Disord 26:928–937
42. Mei H, Chen W, Srinivasan SR, Jiang F, Schork N, Murray S, Smith E, So JD, Berenson GS (2010) FTO influences on longitudinal BMI over childhood and adulthood and modulation on relationship between birth weight and longitudinal BMI. Hum Genet 128:589–96
43. Ross R (1999) Atherosclerosis–an inflammatory disease. N Engl J Med 340:115–126

44. Pearson TA, Mensah GA, Alexander RW, Anderson JL, Cannon RO 3rd, Criqui M, Fadl YY, Fortmann SP, Hong Y, Myers GL, Rifai N, Smith SC Jr, Taubert K, Tracy RP, Vinicor F, Centers for Disease Control and Prevention, American Heart Association (2003) Markers of inflammation and cardiovascular disease: application to clinical and public health practice: a statement for healthcare professionals from the Centers for Disease Control and Prevention and the American Heart Association. Circulation 107:499–511
45. Grundy SM (2003) Inflammation, hypertension, and the metabolic syndrome. J Am Med Assoc 290:3000–3002
46. Trevisanuto D, Doglioni N, Altinier S, Zaninotto M, Plebani M, Zanardo V (2007) High-sensitivity C-reactive protein in umbilical cord of small-for-gestational-age neonates. Neonatology 91:186–189
47. Sattar N, McConnachie A, O'Reilly D, Upton MN, Greer IA, Davey Smith G, Watt G (2004) Inverse association between birth weight and C-reactive protein concentrations in the MIDSPAN Family Study. Arterioscler Thromb Vasc Biol 24:583–587
48. Ford ES (2003) The metabolic syndrome and C-reactive protein, fibrinogen, and leukocyte count: findings from the Third National Health and Nutrition Examination Survey. Atherosclerosis 168:351–358
49. Madjid M, Awan I, Willerson JT, Casscells SW (2004) Leukocyte count and coronary heart disease: implications for risk assessment. J Am Coll Cardiol 44:1945–1956
50. Baumgartner RN, Roche AF, Guo S, Lohman T, Boileau RA, Slaughter MH (1986) Adipose tissue distribution: the stability of principal components by sex, ethnicity and maturation stage. Hum Biol 58:719–735
51. Caprio S, Plewe G, Diamond MP, Simonson DC, Boulware SD, Sherwin RS, Tamborlane WV (1989) Increased insulin secretion in puberty: a compensatory response to reductions in insulin sensitivity. J Pediatr 114:963–967
52. Jiang X, Srinivasan SR, Radhakrishnamurthy B, Dalferes ER, Berenson GS (1996) Racial (black–white) differences in insulin secretion and clearance in adolescents: the Bogalusa Heart Study. Pediatrics 97:357–360
53. Bhuiyan AR, Li S, Li H, Chen W, Srinivasan SR, Berenson GS (2005) Distribution and correlates of arterial compliance measures in asymptomatic young adults: the Bogalusa Heart Study. Am J Hypertens 18:684–691
54. Cohn JN (2006) Arterial stiffness, vascular disease, and risk of cardiovascular events. Circulation 113:601–603
55. Blacher J, Asmar R, Djane S, London GM, Safar ME (1999) Aortic pulse wave velocity as a marker of cardiovascular risk in hypertensive patients. Hypertension 33:1111–1117
56. Finkelstein SM, Cohn JN (1992) First- and third-order models for determining arterial compliance. J Hypertens 10 (Suppl 6):S11–S14
57. te Velde SJ, Ferreira I, Twisk JW, Stehouwer CD, van Mechelen W, Kemper HC (2004) Amsterdam Growth and Health Longitudinal Study. Birthweight and arterial stiffness and blood pressure in adulthood–results from the Amsterdam Growth and Health Longitudinal Study. Int J Epidemiol 33:154–161
58. Lurbe E, Torro MI, Carvajal E, Alvarez V, Redón J (2003) Birth weight impacts on wave reflections in children and adolescents. Hypertension 41:646–650
59. Norman M (2008) Low birth weight and the developing vascular tree: a systematic review. Acta Paediatr 97:1165–1172
60. Broyd C, Harrison E, Raja M, Millasseau SC, Poston L, Chowienczyk PJ (2005) Association of pulse waveform characteristics with birth weight in young adults. J Hypertens 23:1391–1396
61. Montgomery AA, Ben-Shlomo Y, McCarthy A, Davies D, Elwood P, Smith GD (2000) Birth size and arterial compliance in young adults. Lancet 355:2136–2137
62. Martyn CN, Barker DJ, Jespersen S, Greenwald S, Osmond C, Berry C (1995) Growth in utero, adult blood pressure, and arterial compliance. Br Heart J 73:116–121
63. Liao D, Arnett DK, Tyroler HA, Riley WA, Chambless LE, Szklo M, Heiss G (1999) Arterial stiffness and the development of hypertension. The ARIC study. Hypertension 34:201–206

64. Li S, Chen W, Srinivasan SR, Berenson GS (2004) Childhood blood pressure as a predictor of arterial stiffness in young adults: the Bogalusa Heart Study. Hypertension 43:541–546
65. Eriksson JG (2002) Growth and coronary heart disease in adult life. CVR & R 23:557–560
66. Phillips DIW (1996) Insulin resistance as a programmed response to fetal malnutrition. Diabetologia 39:1119–1122
67. Taylor DJ, Thompson CH, Kemp GJ, Barnes PR, Sanderson AL, Radda GK, Phillips DI (1995) A relationship between impaired fetal growth and reduced muscle glycolysis revealed by P-31 magnetic resonance spectroscopy. Diabetologia 38:1205–1212
68. Lapillonne A, Braillon P, Claris O, Chatelain PG, Delmas PD, Salle BL (1997) Body composition in appropriate and in small for gestational age infants. Acta Paediatr 86:196–200
69. Hales CN, Barker DJ (1992) Type 2 (non-insulin-dependent) diabetes mellitus: the thrifty phenotype hypothesis. Diabetologia 35:595–601
70. Dahri S, Reusens B, Remacle C, Hoet JJ (1995) Nutritional influences on pancreatic development and potential links with non-insulin-dependent diabetes. Proc Nutr Soc 54:345–356
71. Barker DJ, Martyn CN, Osmond C, Wield GA (1995) Abnormal liver growth in utero and death from coronary heart disease. Br Med J 310:703–704
72. Mackenzie HS, Brenner BM (1995) Fewer nephrons at birth: a missing link in the etiology of essential hypertension? Am J Kidney Dis 26:91–98
73. Hinchliffe SA, Lynch MR, Sargent PH, Howard CV, Van Velzen D (1992) The effect of intrauterine growth retardation on the development of renal nephrons. Br J Obstet Gynaecol 99:296–301
74. Phillips DIW, Barker DJP (1997) Association between low birthweight and high resting pulse in adult life: is the sympathetic nervous system involved in programming the insulin resistance syndrome? Diabet Med 14:673–677
75. Nuyt AM, Alexander BT (2009) Developmental programming and hypertension. Curr Opin Nephrol Hypertens 18:144–152
76. Phillips DI, Barker DJ, Fall CH, Seckl JR, Whorwood CB, Wood PJ, Walker BR (1998) Elevated plasma cortisol concentrations: a link between low birth weight and the insulin resistance syndrome? J Clin Endocrinol Metab 83:757–760
77. Leeson CP, Whincup PH, Cook DG, Donald AE, Papacosta O, Lucas A, Deanfield JE (1997) Flow-mediated dilation in 9- to 11-year-old children: the influence of intrauterine and childhood factors. Circulation 96:2233–2238
78. Widdowson EM, Crabb DE, Milner RD (1972) Cellular development of some human organs before birth. Arch Dis Child 47:652–655
79. Gustafson B, Hammarstedt A, Andersson CX, Smith U (2007) Inflamed adipose tissue: a culprit underlying the metabolic syndrome and atherosclerosis. Arterioscler Thromb Vasc Biol 27:2276–2283
80. Martyn CN, Greenwald SE (1997) Impaired synthesis of elastin in walls of aorta and large conduit arteries during early development as an initiating event in pathogenesis of systemic hypertension. Lancet 350:953–955
81. Cheung YF, Wong KY, Lam BC, Tsoi NS (2004) Relation of arterial stiffness with gestational age and birth weight. Arch Dis Child 89:217–221

Chapter 3
Trajectories of Variables Related to Cardio-Metabolic Risk from Childhood to Young Adulthood

Sathanur R. Srinivasan and JiHua Xu

Abstract The Framingham Study which coined the term "risk factors" in middle-aged and older populations, provided the rationale for bridging the gap from youth to aging adults as part of the evolution of early natural history of cardiovascular (CV) diseases. This report details the time-course or trajectories of cardio-metabolic risk variables from childhood to adulthood in the Bogalusa Heart Study cohort. The observed adverse trajectories of body fatness, metabolic, and hemodynamic variables since childhood and the potential underlying mechanisms governing their interrelationships support a primary role of excess adiposity in the early natural histories of CV diseases, type 2 diabetes, and hypertension. These findings have implications for public health approaches to prevention.

Keywords Cardio-metabolic risk • Childhood • CV risk factor • Hypertension • Metabolic syndrome • Trajectories • Type 2 diabetes • Younger adulthood

3.1 Introduction

The Framingham Study developed the concept of risk factors in predicting future cardiovascular (CV) risk in middle-aged and older populations [1]. Kannel et al. [2] emphasized the need for bridging the gap from youth to aging adults in terms of trends in the burden of predisposing CV risk factors with age. Autopsy studies from the Bogalusa Heart Study and the Pathobiologic Determinants of Atherosclerosis in Youth Study provided irrefutable anatomic evidence of early "silent" subclinical disease in the CV system in relation to CV risk factor in young individuals [3–6]. Consequently, the view that the pathophysiologic precursors of CV diseases and their risk factors begin in childhood is now well recognized. This complex interplay of the burden of CV risk factors vis-à-vis growth and development and the aging

S. R. Srinivasan (✉)
Departments of Epidemiology and Biochemistry, Center for Cardiovascular Health, Tulane University School of Public Health and Tropical Medicine, New Orleans, LA, USA
e-mail: ssriniv1@tulane.edu

G. S. Berenson (ed.), *Evolution of Cardio-Metabolic Risk from Birth to Middle Age*,
DOI 10.1007/978-94-007-1451-9_3, © Springer Science+Business Media B.V. 2011

process has its' origin even in utero [7]. Importantly, dynamic changes in interrelated risk variables such as adiposity, lipoproteins, hemodynamic variables, and measure of glucose homeostasis that occur during growth and maturation in childhood and beyond potentially influence cardio-metabolic risk in terms of morbidity and mortality. In this regard, longitudinal epidemiologic studies from childhood to adulthood provide information on the evolution of natural history of CV diseases [8–11]. The intent of this chapter is to provide time-course or trajectories of variables related to cardio-metabolic risk since childhood based on observations made in a longitudinal biracial (black–white) cohort enrolled in the Bogalusa Heart Study [12–17].

3.2 Cardiovascular Disease

In younger adults, it is not feasible, for obvious reasons, to determine the relationship between longitudinal changes in risk factor variables since childhood and incident CV disease morbidity and mortality in adulthood. Parental history of CV disease provides a viable means to examine this relationship, since CV diseases aggregate in families [18, 19], and parental history is an established surrogate measure of future CV diseases in offspring [20].

With respect to lipoprotein variables (Fig. 3.1), the longitudinal trends over time, beginning during childhood and extending into adulthood, the offspring with

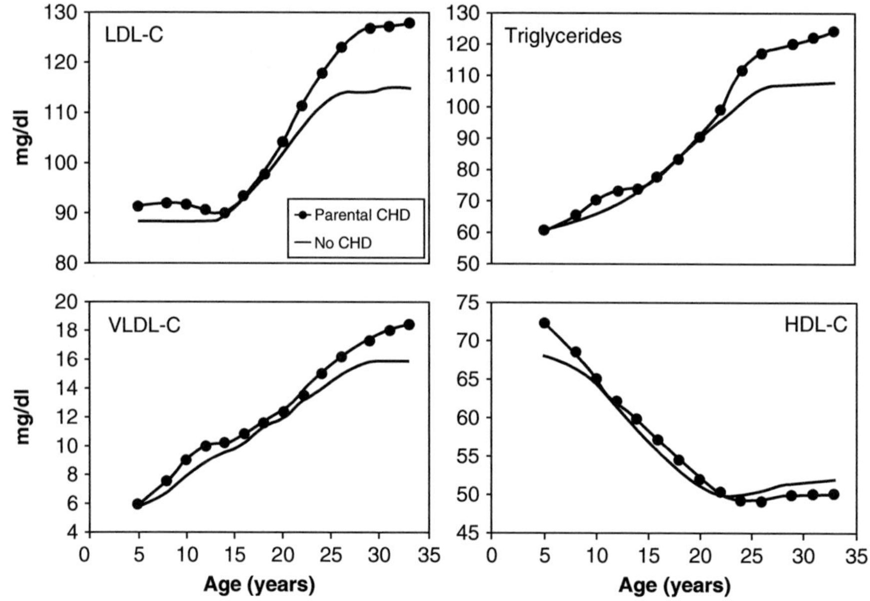

Fig. 3.1 Longitudinal trends in serum lipoprotein variables in the offspring from childhood to adulthood by parental CV disease status as determined by generalized estimation equations: the Bogalusa Heart Study. [13]

parental history of CV disease showed accentuated levels of serum low-density lipoprotein cholesterol (LDL-C) and triglycerides after their maturation into adulthood; a small but consistent higher levels of very-low-density lipoprotein cholesterol (VLDL-C) during both childhood and young adulthood. An inverse trend of high-density-lipoprotein cholesterol (HDL-C) with age was noted between the ages 5 and 20 years in both groups, although at later ages a slight trend for lower values was seen for the offspring with parental history. All longitudinal trends of lipoprotein variables by parental history of CV disease status were consistent, regardless of race and sex. The adverse trends in lipoprotein variables, especially LDL-C and triglycerides, in the offspring of parents with CV disease were accentuated more in adulthood than during childhood, probably reflecting the burden of interaction between genes and lifestyle-related factors with age. It is of interest that parental CV disease was no longer a predictor of adverse longitudinal trends in lipoprotein variables in the offspring when adjusted for measures of body fatness and variables of glucose homeostasis (fasting glucose and insulin). This is consistent with the well recognized interrelationships among these variables [21].

With respect to obesity, its association with CV diseases, type 2 diabetes mellitus, and hypertension is thought to be mediated through the development of insulin resistance/hyperinsulinemia and the attendant adverse changes in lipoproteins, blood pressure, and glucose tolerance [21–23]. The interrelationships or clustering of these cardio-metabolic risk factor variables is found to persist from childhood to adulthood in the Bogalusa Heart Study cohort [24]. Further, as shown in Fig. 3.2, the time-course relations of parenteral history of CV disease with body fatness and fasting insulin in the offspring from childhood to younger adulthood indicate that the affected offspring had excess generalized and truncal adiposity already in childhood, independent of age, race, and sex; this trend persisted into younger adulthood [14]. On the other hand, higher fasting insulin levels among those with affected parents became evident only after age 20, after completion of growth and sexual maturation. This may be the manifestation of long standing burden of obesity beginning in childhood.

Although the association between obesity and hyperinsulinemia/insulin resistance is well known, studies have shown temporal sequence of obesity and hyperinsulinemia to occur in both directions [25, 26]. A number of putative mechanisms that underlie this temporal sequence either way have been suggested [27, 28]. It has been hypothesized that populations genetically predisposed to obesity and type 2 diabetes are endowed with the thrifty genotype, which was intended to facilitate efficient fat storage in times of food abundance through a high insulin response to provide an energy buffer in times of scarcity [29]. When a rapid transition occurs from a traditional subsistence lifestyle to a western lifestyle, hyperinsulinemia may precede obesity, as in the case of Pima Indians [26]. However, the thrifty genotype metabolic sequelae may not be applicable to the U.S. general population. If fact, studies on the Bogalusa Heart Study cohort found a significant positive association between the degree of baseline obesity and the incidence of hyperinsulinemia at follow-up in children, adolescents, and young adults alike, independent of baseline

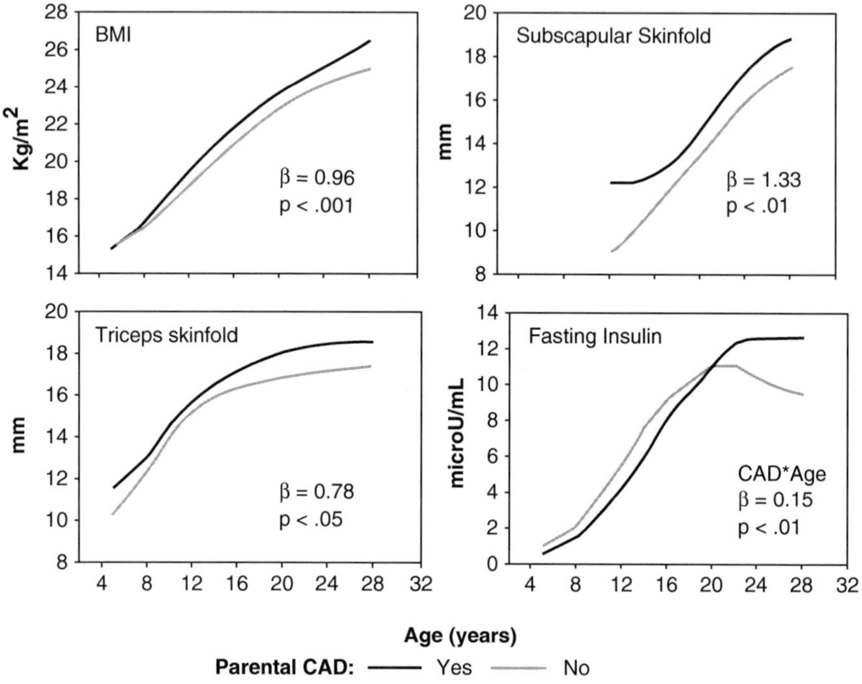

Fig. 3.2 Longitudinal trends in body fatness and fasting insulin in the offspring from childhood to adulthood by parental CV disease status as determined by generalized estimation equations: the Bogalusa Heart Study. [14]

insulin, race and sex (Fig. 3.3), thereby supporting the role of obesity in the development of hyperinsulinemia/insulin resistance beginning in childhood [30, 31].

3.3 Diabetes Mellitus

Type 2 diabetes, one of the most commonly prevalent chronic diseases in the U.S, increases the risk for CV morbidity and premature mortality [32]. Of concern is the increase in the incidence of type 2 diabetes among children and adolescents over the past decade [33, 34]. It is widely recognized that type 2 diabetes is preceded by a long prediabetes stage characterized by metabolic abnormalities including insulin resistance [35–37]. Both type 2 diabetes and CV diseases are associated with a constellation of disorders characteristic of metabolic syndrome [21–23]. Of interest, childhood observations in the Bogalusa cohort found black adolescents having higher insulin and lower C-peptide levels than their white counterparts, indicating lower insulin secretion by the beta cells in black adolescents [38]. Moreover, black adolescents had lower C-peptide to insulin ratio than white adolescents, suggesting reduced hepatic clearance in black adolescents. Also, the observed significantly lower

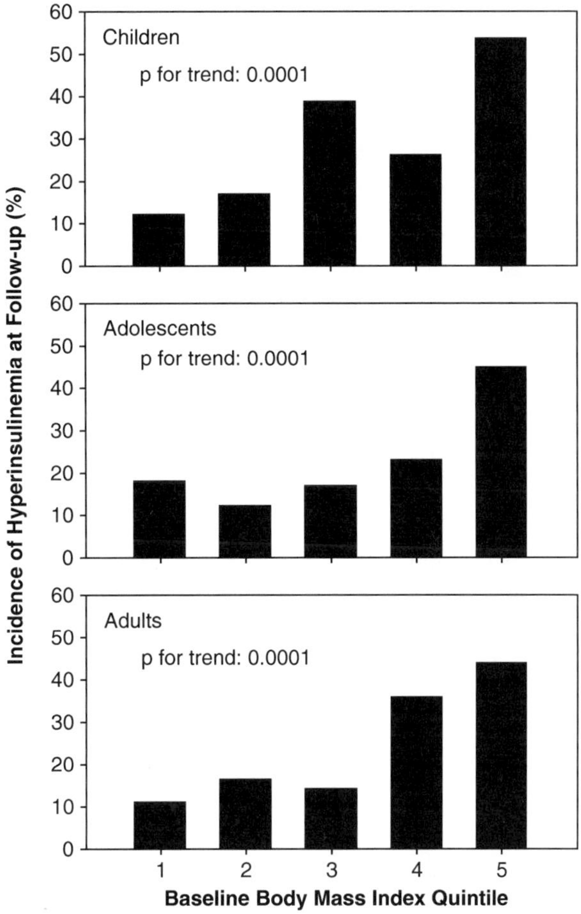

Fig. 3.3 Incidence of hyperinsulinemia (insulin>75th percentile specific for age, race, sex and survey year) at follow-up study in children, adolescents, and adults by body mass index quintiles (specific for age, race, sex and survey year) at baseline: the Bogalusa Heart Study. [30]

levels of glucose to insulin ratio in black girls suggests a reduced insulin sensitivity in this group. With respect to obesity, both increased insulin secretion and decreased insulin clearance contributed to hyperinsulinemia in obese adolescents [39].

Longitudinal traits of risk variables of metabolic syndrome since childhood by diabetes status in adulthood, as illustrated in Fig. 3.4, revealed consistently higher levels of age-, race-, and sex-adjusted fasting glucose from childhood through adulthood in both prediabetic and diabetic vs normoglycemics subjects; higher LDL-C, insulin and insulin resistance index in prediabetics since adolescence; and higher body fatness, triglycerides, insulin, and insulin resistance index (HOMA-IR), and lower HDL-C among diabetics since childhood [16]. The rate of increase in body fatness, triglycerides, glucose, insulin and insulin resistance index were significantly higher among prediabetics and diabetics; mean arterial pressure among diabetics. Adverse longitudinal changes of glucose, insulin resistance index, and LDL-C from childhood to adulthood were independently related to prediabetes status; body

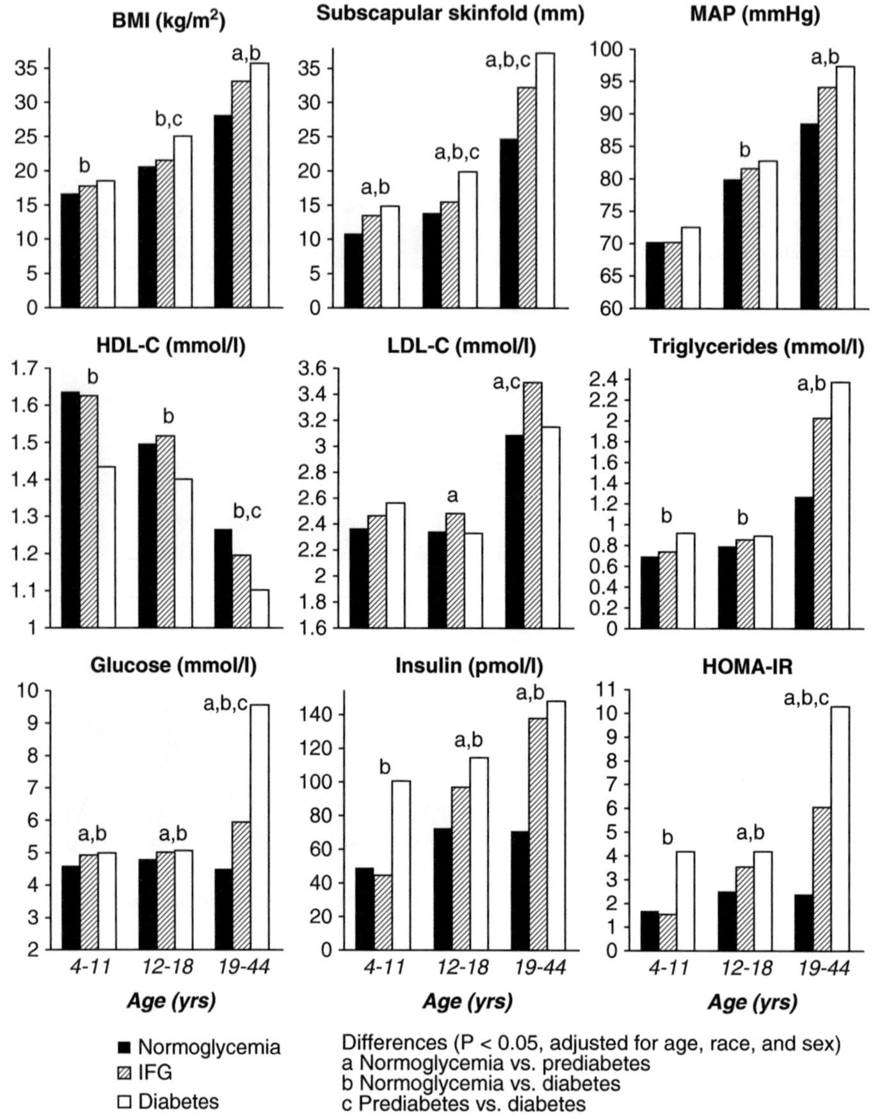

Fig. 3.4 Mean levels of body mass index, subscapular skinfold, mean arterial pressure, HDL-C, LDL-C, triglycerides, fasting glucose, insulin and insulin resistance index (HOMA-IR) from childhood to adulthood by adult diabetes status: the Bogalusa Heart Study. [16]

fatness, HDL-C, and insulin resistance index to diabetes status. In terms of prevalence, the observed differences of sex (males > females) in the prediabetic group and race (blacks > whites) in the diabetic group in this cohort are consistent with earlier reports [40, 41]. Also, as in previous studies [40, 41], prediabetes and diabetes were less prevalent among white females.

The observation that excess adiposity occurred beginning in childhood, changed adversely through adulthood, and related independently with diabetes in adulthood in the Bogalusa Heart Study cohort is also noted in other population studies [42, 43]. Further, in addition to impaired insulin sensitivity, defect in insulin-independent glucose utilization (glucose effectiveness) is considered an important trait for developing diabetes [44]. The adverse longitudinal profiles of baseline insulin and insulin resistance index in conjunction with increased fasting glucose among diabetes seen in the Bogalusa cohort also suggest slow glucose removal beginning in childhood.

Importantly, the prevalent rate of adult diabetes status by quartiles of baseline childhood fasting glucose levels in the Bogalusa cohort showed significant adverse trend for both prediabetes and diabetes (Fig. 3.5), with an apparent threshold occurring at or above the 50th percentile (86 mg/dL), which is well within the currently accepted reference range and consistent with earlier studies [45, 46]. Regarding the predictive value of the above threshold, the area under the receiver operating curve analysis yielded a C value of 0.86 for prediabetes and 0.79 for diabetes, with sensitivity and specificity, respectively, of 77 and 85% for prediabetes and 75% and 76% for diabetes. In a multivariate analysis that included anthropometric, metabolic and hemodynamic variables from childhood to adulthood along with baseline childhood fasting glucose status (< 50th percentile vs the rest), individuals with elevated childhood glucose levels were 3.4 and 2.1 times more likely to develop prediabetes and diabetes, respectively, as adults. In addition, consistent with earlier studies [43, 45, 47], adverse levels of LDL-C and insulin resistance index in childhood also emerged as independent predictors of prediabetes; obesity measures and insulin resistance index of diabetes. The prevalence of metabolic syndrome in the Bogalusa cohort was higher in adulthood among those with prediabetes and diabetes, as might be expected [48]. Thus, adverse levels of risk variables of metabolic syndrome, adiposity and measures of glucose homeostasis in particular, and their accelerated rates of change since childhood characterize the early natural history of type 2 diabetes.

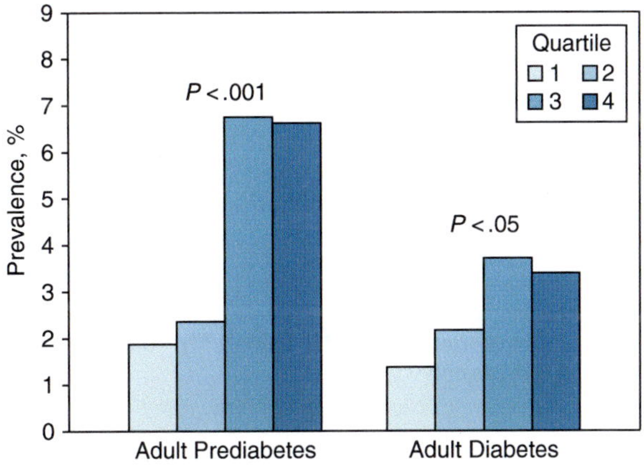

Fig. 3.5 Prevalence of adult diabetes status by quartiles of childhood fasting glucose levels within the normoglycemics range. Levels of fasting glucose according to the quartiles were less than 80 mg/dL for quartile1; 80–85 mg/dL for quartile2; 86–90 mg/dL for quartile3; and 91–99 mg/dL for quartile4: the Bogalusa Heart Study

3.4 Hypertension

Essential hypertension is an important risk factor for morbidity and mortality from coronary heart disease, stroke, and renal disease [49, 50]. Although clinical manifestations of hypertension do not generally emerge until middle age, the pathobiologic precursors of adult hypertension are thought to originate very early in life, including the period of fetal development [9–11, 51].

It is now well recognized that hypertension, in general, does not occur in isolation but co-exists in varying degrees with disorders characteristics of metabolic syndrome [21–23]. Further, prospective studies have shown that hypertension is preceded by a prehypertension stage characterized by abnormalities considered as potential metabolic precursors of hypertension [52–54]. The Bogalusa Heart Study focused on the course of concurrent development of adverse levels of blood pressure and other risk variables of metabolic syndrome during childhood (4–11 years), adolescence (12–18 years) and adulthood (19–42 years) in a longitudinal cohort [17].

Prehypertensive vs normotensive subjects had significantly higher measures of adiposity, systolic and diastolic blood pressures, and triglycerides beginning in childhood; higher glucose in adolescence; and higher LDL-C, fasting insulin, and insulin resistance index in adulthood. Hypertensive vs normotensive subjects displayed higher adiposity measures, systolic and diastolic blood pressures, fasting glucose, and triglycerides beginning in childhood; higher fasting insulin and insulin resistance index in childhood and adulthood; and lower HDL-C in adulthood. As shown in Table 3.1, most of these variables progressed adversely over time at an increased rate in both prehypertensive and hypertensive subjects. Adverse changes in systolic and diastolic blood pressures and adiposity were inde-

Table 3.1 Longitudinal rates of change in risk variables of metabolic syndrome since childhood in the study cohort by adult hypertension status: the Bogalusa Heart Study. [17]

Variable	Normotensive	Prehypertensive	Hypertensive
BMI (kg/m² per year)	0.42	0.55*	0.56*
Subscapular skinfold (mm/year)	0.68	0.81*	0.92*
Systolic BP (mm Hg/year)	0.36	0.87*	0.96*
Diastolic BP (mm Hg/year)	0.50	0.84*	0.92*
LDL cholesterol (mg/dL per year)	1.42	1.93*	1.80*
HDL cholesterol (mg/dL per year)	−0.52	−0.59	−0.49
Triglycerides (mg/dL per year)[a]	2.08	3.32*	4.52*
Insulin (μU/mL per year)[a]	0.05	0.16***	0.16†
Glucose (mg/dL per year)[a]	−0.74	−0.83	−0.54
Insulin resistance index (year)[a]	−0.02	0.01**	0.01†

Values are regression slope with respect to age in years adjusted for race and sex, and the race by sex interaction, as applicable

Different from normotensives: *$p<0.0001$; **$p<0.001$; ***$p<0.01$; † $p<0.05$

[a] Fasting subjects only

pendently related to prehypertensive status; adiposity, systolic and diastolic blood pressures, insulin resistance index, LDL-C, HDL-C, and triglycerides to hypertension status. As young adults, prehypertensive and hypertensive subjects had significantly higher prevalence of obesity, hyperinsulinemia, hyperglycemia and dyslipidemia.

The findings that blood pressure levels were higher beginning in childhood, changed adversely through adulthood, and associated independently with conditions of prehypertension and hypertension in adulthood is consistent with the concept of tracking (persistence) of risk factor variables over time. Persistence of elevated blood pressure over time has been demonstrated in both pediatric and adult populations [55–57]. Studies have also shown that earlier blood pressure elevations ultimately progress to clinical hypertension [55–59]. The clinical significance of childhood elevations in blood pressure progressing to adult hypertension is underscored by the fact that adverse anatomic CV changes characteristics of hypertension already occur in children with elevated blood pressure [60–63]. Further, in the Bogalusa cohort study [64], childhood blood pressure and its rate of change from childhood to adulthood were significant predictors of microalbuminuria, a biomarker of future development of diabetic nephropathy, renal disease, and CV mortality [65], in blacks, but not in whites.

The precise mechanisms linking the antecedent factors to hypertension are uncertain at present. That both prehypertensive and hypertensive Bogalusa cohort displayed excess adiposity since childhood support earlier studies showing obesity as an independent risk factor for hypertension [53, 56, 66]. Obesity, per se, could raise blood pressure by adversely altering intravascular volume, cardiac output, cardiac systolic and diastolic functions, renal-pressure natriuresis, and renal medullary compression [67, 68]. As a highly active endocrine organ, adipose tissue also plays an important role in the regulation of metabolic and hemodynamic processes through mechanisms that include activation of sympathetic nervous system and adipose renin-angiotensin-aldosterone system along with suppression of natriuretic peptides activity [69–72]. With respect to hyperinsulinemia/insulin resistance, its association with hypertension may reflect concurrent mechanisms related to the effects of excess adiposity on blood pressure and insulin [72, 73].

3.5 Conclusion

The adverse trajectories of anthropometric, metabolic and hemodynamic variables since childhood in the Bogalusa Heart Study cohort support a primary role for excess adiposity in the early natural histories of CV diseases, type 2 diabetes, and hypertension. In terms of public health, when viewed in the context of upward secular trend in adiposity among youth [74, 75], these findings underscore the importance of primordial prevention.

References

1. Kannel WB (1988) Contributions of the Framingham Study to the conquest of coronary artery disease. Am J Cardiol 62:1109–1112
2. Kannel WB, D'Agostino RB, Belanger AJ (1995) Concept of bridging the gap from youth to adulthood. Am J Med Sci 310 (Suppl 1):S15–S21
3. Newman WP III, Freedman DS, Voors AW, Gard PD, Srinivasan SR, Cresanta JL, Williamson GD, Webber LS, Berenson GS (1986) Relation of serum lipoprotein levels and systolic blood pressure to early atherosclerosis. The Bogalusa Heart Study. N Engl J Med 314:138–144
4. Pathobiological Determinants of Atherosclerosis in Youth (PDAY) (1990) Research group: relationship of atherosclerosis in young men to serum lipoprotein cholesterol concentrations and smoking. J Am Med Assoc 264:3018–3024
5. Berenson GS, Srinivasan SR, Bao W, Newman WP III, Tracy RE, Wattigney WA (1998) Association between multiple cardiovascular risk factors and atherosclerosis in children and young adults. The Bogalusa Heart Study. N Engl J Med 338:1650–1656
6. Napoli C, Glass CK, Witztum JL, Deutsch R, D'Armiento FP, Palinski W (1999) Influence of maternal hypercholesterolaemia during pregnancy on progression of early atherosclerotic lesions in childhood: Fate of Early Lesions in Children (FELIC) study. Lancet 354:1234–1241
7. Barker DJ (1994) Programming the baby. In: Barker DJ (ed) Mothers, babies, and disease later in Life. BMJ Publishing Group, London, pp 14–36
8. Berenson GS, McMahan CA, Voors AW, Webber LS, Srinivasan SR, Frank GC, Foster TA, Blonde CV (1980) Cardiovascular risk factors in children. In: Andrews C, Hester HE (eds) The early natural history of atherosclerosis and essential hypertension. Oxford University Press, New York, pp 47–123
9. Berenson GS (ed) (1986) Causation of cardiovascular risk factors in children: perspectives on cardiovascular risk in early life. Raven Press, New York, pp 1–408
10. Lauer RM, Shekelle RB (eds) (1980) Childhood prevention of atherosclerosis and hypertension. Raven Press, New York, pp 1–484
11. Lauer RM, Burns TL, Daniels SR (eds) (2006) Pediatric prevention of atherosclerotic cardiovascular disease. Oxford University Press, New York, pp 1–378
12. Bao W, Srinivasan SR, Valdez R, Greenlund KJ, Wattigney WA, Berenson GS (1997) Longitudinal changes in cardiovascular risk from childhood to young adulthood in offspring of parents with coronary artery disease: the Bogalusa Heart Study. J Am Med Assoc 278:1749–1754
13. Youssef AA, Srinivasan SR, Elkasabany A, Chen W, Berenson GS (2001) Trends of lipoprotein variables from childhood to adulthood in offspring of parents with coronary heart disease: the Bogalusa Heart Study. Metabolism 50:1441–1446
14. Youssef AA, Valdez R, Elkasabany A, Srinivasan SR, Berenson GS (2002) Time-course of adiposity and fasting insulin from childhood to young adulthood in offspring of parents with coronary artery disease: the Bogalusa Heart Study. Ann Epidemiol 12:553–559
15. Srinivasan SR, Frontini MG, Berenson GS (2003) Bogalusa Heart Study. Longitudinal changes in risk variables of insulin resistance syndrome from childhood to young adulthood in offspring of parents with type 2 diabetes: the Bogalusa Heart Study. Metabolism 52:443–450
16. Nguyen QM, Srinivasan SR, Xu JH, Chen W, Berenson GS (2008) Changes in risk variables of metabolic syndrome since childhood in pre-diabetic and type 2 diabetic subjects: the Bogalusa Heart Study. Diabetes Care 31:2044–2049
17. Srinivasan SR, Myers L, Berenson GS (2006) Changes in metabolic syndrome variables since childhood in prehypertensive and hypertensive subjects: the Bogalusa Heart Study. Hypertension 48:33–39
18. Barrett-Connor E, Khaw K (1984) Family history of heart attack as an independent predictor of death due to cardiovascular disease. Circulation 69:1065–1069

19. ten Kate LP, Boman H, Daiger SP, Motulsky AG (1982) Familial aggregation of coronary heart disease and its relation to known genetic risk factors. Am J Cardiol 50:945–953
20. Summary of the Second Report of the National Cholesterol Education Program (NCEP) (1993) Expert panel on detection, evaluation, and treatment of high blood cholesterol in adults (Adult Treatment Panel II). J Am Med Assoc 269:3015–3023
21. Reaven GM (1988) Banting lecture 1988. Role of insulin resistance in human disease. Diabetes 37:1595–1607
22. DeFronzo RA, Ferrannini E (1991) Insulin resistance. A multifaceted syndrome responsible for NIDDM, obesity, hypertension, dyslipidemia, and atherosclerotic cardiovascular disease. Diabetes Care 14:173–194
23. Haffner SM, Valdez RA, Hazuda HP, Mitchell BD, Morales PA, Stern MP (1992) Prospective analysis of the insulin-resistance syndrome (syndrome X). Diabetes 41:715–722
24. Bao W, Srinivasan SR, Wattigney WA, Berenson GS (1994) Persistence of multiple cardiovascular risk clustering related to syndrome X from childhood to young adulthood. The Bogalusa Heart Study. Arch Intern Med 154:1842–1847
25. Lazarus R, Sparrow D, Weiss S (1998) Temporal relations between obesity and insulin: longitudinal data from the Normative Aging Study. Am J Epidemiol 147:173–179
26. Odeleye OE, de Courten M, Pettitt DJ, Ravussin E (1997) Fasting hyperinsulinemia is a predictor of increased body weight gain and obesity in Pima Indian children. Diabetes 46:1341–1345
27. Ravussin E, Swinburn BA (1996) Insulin resistance is a result, not a cause of obesity. Socratic debate: the pro side. In: Angel A, Anderson H, Bouchard C, Lau D, Leiter L, Mendelson R (eds) Progress in obesity research. Seventh International Congress in Obesity. Libbey, London, pp 173–178
28. Sims EAH (1996) Insulin resistance is a result, not a cause of obesity. Socratic debate: the con side. In: Angel A, Anderson H, Bouchard C, Lau D, Leiter L, Mendelson R (eds) Progress in obesity research. Seventh International Congress in Obesity. Libbey, London, pp 587–592
29. Neel JV (1962) Diabetes mellitus: a "thrifty" genotype rendered detrimental by progress? Am J Hum Genet 14:353–362
30. Srinivasan SR, Myers L, Berenson GS (1999) Temporal association between obesity and hyperinsulinemia in children, adolescents, and young adults: the Bogalusa Heart Study. Metabolism 48:928–934
31. Sinaiko AR, Donahue RP, Jacobs DR Jr, Prineas RJ (1999) Relation of weight and rate of increase in weight during childhood and adolescence to body size, blood pressure, fasting insulin, and lipids in young adults. The Minneapolis Children's Blood Pressure Study. Circulation 99:1471–1476
32. Fox CS, Sullivan L, D'Agostino RB Sr, Wilson PW (2004) Framingham Heart Study. The significant effect of diabetes duration on coronary heart disease mortality: the Framingham Heart Study. Diabetes Care 27:704–708
33. Pinhas-Hamiel O, Dolan LM, Daniels SR, Standiford D, Khoury PR, Zeitler P (1996) Increased incidence of non-insulin-dependent diabetes mellitus among adolescents. J Pediatr 128:608–615
34. Fagot-Campagna A, Pettitt DJ, Engelgau MM, Burrows NR, Geiss LS, Valdez R, Beckles GL, Saaddine J, Gregg EW, Williamson DF, Narayan KM (2000) Type 2 diabetes among North American children and adolescents: an epidemiologic review and a public health perspective. J Pediatr 136:664–672
35. Warram JH, Martin BC, Krolewski AS, Soeldner JS, Kahn CR (1990) Slow glucose removal rate and hyperinsulinemia precede the development of type II diabetes in the offspring of diabetic parents. Ann Intern Med 113:909–915
36. McCance DR, Pettitt DJ, Hanson RL, Jacobsson LT, Bennett PH, Knowler WC (1994) Glucose, insulin concentrations and obesity in childhood and adolescence as predictors of NIDDM. Diabetologia 37:617–623

37. Haffner SM, Stern MP, Mitchell BD, Hazuda HP, Patterson JK (1990) Incidence of type II diabetes in Mexican Americans predicted by fasting insulin and glucose levels, obesity, and body-fat distribution. Diabetes 39:283–288
38. Jiang X, Srinivasan SR, Radhakrishnamurthy B, Dalferes E Jr, Berenson GS (1996) Racial (black–white) differences in insulin secretion and clearance in adolescents: the Bogalusa Heart Study. Pediatrics 97:357–360
39. Jiang X, Srinivasan SR, Berenson GS (1996) Relation of obesity to insulin secretion and clearance in adolescents: the Bogalusa Heart Study. Int J Obesity 20:951–956
40. Cowie CC, Rust KF, Byrd-Holt DD, Eberhardt MS, Flegal KM, Engelgau MM, Saydah SH, Williams DE, Geiss LS, Gregg EW (2006) Prevalence of diabetes and impaired fasting glucose in adults in the U.S. population: National Health and Nutrition Examination Survey 1999–2002. Diabetes Care 29:1263–1268
41. Pankow JS, Kwan DK, Duncan BB, Schmidt MI, Couper DJ, Golden S, Ballantyne CM (2007) Cardio-metabolic risk in impaired fasting glucose and impaired glucose tolerance: the Atherosclerosis Risk in Communities Study. Diabetes Care 30:325–331
42. Cheung YB, Machin D, Karlberg J, Khoo KS (2004) A longitudinal study of pediatric body mass index values predicted health in middle age. J Clin Epidemiol 57:1316–1322
43. Franks PW, Hanson RL, Knowler WC, Moffett C, Enos G, Infante AM, Krakoff J, Looker HC (2007) Childhood predictors of young-onset type 2 diabetes. Diabetes 56:2964–2972
44. Martin BC, Warram JH, Krolewski AS, Bergman RN, Soeldner JS, Kahn CR (1992) Role of glucose and insulin resistance in development of type 2 diabetes mellitus: results of a 25-year follow-up study. Lancet 340:925–929
45. Tirosh A, Shai I, Tekes-Manova D, Israeli E, Pereg D, Shochat T, Kochba I, Rudich A (2005) Israeli Diabetes Research Group. Normal fasting plasma glucose levels and type 2 diabetes in young men. N Engl J Med 353:1454–1462
46. Park YW, Chang Y, Sung KC, Ryu S, Sung E, Kim WS (2006) The sequential changes in the fasting plasma glucose levels within normoglycemic range predict type 2 diabetes in healthy, young men. Diabetes Res Clin Pract 73:329–335
47. Wilson PW, Meigs JB, Sullivan L, Fox CS, Nathan DM, D'Agostino RB Sr (2007) Prediction of incident diabetes mellitus in middle-aged adults: the Framingham Offspring Study. Arch Intern Med 167:1068–1074
48. Nóvoa FJ, Boronat M, Saavedra P, Díaz-Cremades JM, Varillas VF, La Roche F, Alberiche MP, Carrillo A (2005) Differences in cardiovascular risk factors, insulin resistance, and insulin secretion in individuals with normal glucose tolerance and in subjects with impaired glucose regulation: the Telde Study. Diabetes Care 28:2388–2393
49. Stamler J, Stamler R, Neaton JD (1993) Blood pressure, systolic and diastolic, and cardiovascular risks. U.S. population data. Arch Intern Med 153:598–615
50. Vasan RS, Larson MG, Leip EP, Evans JC, O'Donnell CJ, Kannel WB, Levy D (2001) Impact of high-normal blood pressure on the risk of cardiovascular disease. N Engl J Med 345:1291–1297
51. Barker DJ, Osmond C, Golding J, Kuh D, Wadsworth ME (1989) Growth in utero, blood pressure in childhood and adult life, and mortality from cardiovascular disease. Br Med J 298:564–567
52. Haffner SM, Ferrannini E, Hazuda HP, Stern MP (1992) Clustering of cardiovascular risk factors in confirmed prehypertensive individuals. Hypertension 20:38–45
53. Garrison RJ, Kannel WB, Stokes J III, Castelli WP (1987) Incidence and precursors of hypertension in young adults: the Framingham Offspring Study. Prev Med 16:235–251
54. Skarfors ET, Lithell HO, Selinus I (1991) Risk factors for the development of hypertension: a 10-year longitudinal study in middle-aged men. J Hypertens 9:217–223
55. Lauer RM, Clarke WR (1989) Childhood risk factors for high adult blood pressure: the Muscatine Study. Pediatrics 84:633–641
56. Bao W, Threefoot SA, Srinivasan SR, Berenson GS (1995) Essential hypertension predicted by tracking of elevated blood pressure from childhood to adulthood: the Bogalusa Heart Study. Am J Hypertens 8:657–665

57. Beckett LA, Rosner B, Roche AF, Guo S (1992) Serial changes in blood pressure from adolescence into adulthood. Am J Epidemiol 135:1166–1177
58. Kuller LH, Crook M, Almes MJ, Detre K, Reese G, Rutan G (1980) Dormont High School (Pittsburgh, Pennsylvania) blood pressure study. Hypertension 2:109–116
59. Leitschuh M, Cupples LA, Kannel W, Gagnon D, Chobanian A (1991) High-normal blood pressure progression to hypertension in the Framingham Heart Study. Hypertension 17:22–27
60. Schieken RM, Clarke WR, Lauer RM (1981) Left ventricular hypertrophy in children with blood pressures in the upper quintile of the distribution. The Muscatine Study. Hypertension 3:669–675
61. Burke GL, Arcilla RA, Culpepper WS, Webber LS, Chiang YK, Berenson GS (1987) Blood pressure and echocardiographic measures in children: the Bogalusa Heart Study. Circulation 75:106–114
62. Daniels SR, Loggie JM, Khoury P, Kimball TR (1998) Left ventricular geometry and severe left ventricular hypertrophy in children and adolescents with essential hypertension. Circulation 97:1907–1911
63. Sorof JM, Alexandrov AV, Cardwell G, Portman RJ (2003) Carotid artery intimal-medial thickness and left ventricular hypertrophy in children with elevated blood pressure. Pediatrics 111:61–66
64. Hoq S, Chen W, Srinivasan SR, Berenson GS (2002) Childhood blood pressure predicts adult microalbuminuria in African Americans, but not in whites: the Bogalusa Heart Study. Am J Hypertens 15:1036–1041
65. Mogensen CE (1984) Microalbuminuria predicts clinical proteinuria and early mortality in maturity-onset diabetes. N Engl J Med 310:356–360
66. Stamler R, Stamler J, Riedlinger WF, Algera G, Roberts RH (1978) Weight and blood pressure findings in hypertension screening of 1 million Americans. J Am Med Assoc 240:1607–1610
67. Hall JE (2000) Pathophysiology of obesity hypertension. Curr Hypertens Rep 2:139–147
68. Frohlich ED, Messerli FH, Reisin E, Dunn FG (1983) The problem of obesity and hypertension. Hypertension 5:S71–S78
69. Kershaw EE, Flier JS (2004) Adipose tissue as an endocrine organ. J Clin Endocrinol Metab 89:2548–2556
70. Engeli S, Sharma AM (2001) The renin-angiotensin system and natriuretic peptides in obesity-associated hypertension. J Mol Med 79:21–29
71. Zhang R, Reisin E (2000) Obesity-hypertension: the effects on cardiovascular and renal systems. Am J Hypertens 13:1308–1314
72. Reaven GM, Lithell H, Landsberg L (1996) Hypertension and associated metabolic abnormalities—the role of insulin resistance and the sympathoadrenal system. N Engl J Med 334:374–381
73. Hall JE, Brands MW, Zappe DH, Alonso Galicia M (1995) Insulin resistance, hyperinsulinemia, and hypertension: causes, consequences, or merely correlations? Proc Soc Exp Biol Med 208:317–329
74. Ogden CL, Flegal KM, Carroll MD, Johnson CL (2002) Prevalence and trends in overweight among US children and adolescents, 1999–2000. J Am Med Assoc 288:1728–1732
75. Broyles S, Katzmarzyk PT, Srinivasan SR, Chen W, Bouchard C, Freedman DS, Berenson GS (2010) The pediatric obesity epidemic continues unabated in Bogalusa, Louisiana. Pediatrics 125:900–905

Chapter 4
Evolution of Metabolic Syndrome from Childhood

Wei Chen

Abstract Metabolic syndrome (MetS), the concurrence of inextricably linked disorders including obesity, insulin resistance, dyslipidemia and hypertension, has gained importance because of its association with subsequent morbidity and mortality from cardiovascular disease. This chapter reviews observations on the MetS in children and adults from the Bogalusa Heart Study, a biracial (black–white) community-based longitudinal study of the early natural history of cardiovascular disease beginning in childhood. The evolution of the definition, the role of obesity in the pathogenesis of the syndrome, tracking from childhood to adulthood and black–white difference in prevalence of the MetS are described. With respect to clustering analysis of the MetS components, conventional and advanced statistical methodologies, including observed to expected ratio, intraclass correlation, factor analysis and path analysis, are explained in detail. Observations from the Bogalusa Heart Study reinforce recommendations to prevent and modulate the development of the MetS during childhood.

Keywords Metabolic syndrome • Obesity • Black–white difference • Statistical methodology

4.1 Introduction

Cardiovascular (C-V) disease is the major cause of morbidity and mortality in developed countries. During most of the twentieth century, there was considerable effort to understand the underlying pathobiology of the disease and the contributing risk factors, most notably the traditional risk factors of obesity, type 2 diabetes, dyslipidemia and hypertension. It became apparent that multiple risk factors were often present in the same individual. The clustering of these C-V risk factors was first de-

W. Chen (✉)
Department of Epidemiology, Center for Cardiovascular Health, Tulane University School of Public Health and Tropical Medicine, New Orleans, LA, USA
e-mail: wchen1@tulane.edu

G. S. Berenson (ed.), *Evolution of Cardio-Metabolic Risk from Birth to Middle Age*,
DOI 10.1007/978-94-007-1451-9_4, © Springer Science+Business Media B.V. 2011

scribed by Haller et al. [1]. Of note, this phenomenon was also noted early in 1970s [2, 3], and obesity was found to play a crucial role in the clustering of C-V risk factors in children [4] from the Bogalusa Heart Study (BHS), a biracial (black–white) community-based longitudinal study of the early natural history of C-V disease beginning in childhood [5]. Furthermore, the extent of coronary atherosclerosis and vascular disease among adolescents and young adults has been found to increase markedly with the number of C-V risk factors in the BHS and Pathobiological Determinants of Atherosclerosis in Youth (PDAY) autopsy studies [6, 7].

The landmark publication of Reaven's 1988 Banting Medal award lecture highlighted this interrelated phenomenon [8]. Reaven postulated that insulin resistance and its compensatory hyperinsulinemia predisposed patients to hypertension, hyperlipidemia, and termed this phenomenon as "syndrome X". Although obesity was not included in Reaven's primary list of disorders, he acknowledged that it was correlated with insulin resistance or hyperinsulinemia. The condition of this risk factor clustering was also called "insulin resistance syndrome" [9]. Later, the World Health Organization (WHO) [10] and the Third Report of the National Cholesterol Education Program (NCEP) Adult Treatment Panel (ATP III) [11, 12] proposed to use "metabolic syndrome (MetS)" to describe the condition. Although significant clustering of the MetS components has been consistently demonstrated in various ethnic groups and populations, there are still some concerns and debates regarding the definition of the MetS [13]. Several organizations have proposed definitions of the MetS for adults using different components and cut points. One consequence of the nonuniform definition is that currently available data on the frequency of the syndrome in various populations vary widely. A detailed review on the prevalence of the MetS using different criteria has been published recently [14, 15]. In spite of attempts in recent years to reach agreement on the definition of the syndrome, the comparison of published prevalence rates for different populations worldwide is a concern.

4.2 Definition of MetS

The MetS guidelines and definition criteria for adults and children have evolved over the past decade [16–18]. The changes included the concept of the underlying defect, the recommended MetS components and their cut-off values. Insulin resistance was identified as the dominant cause of the MetS in the WHO guidelines in 1998 [19]. As evidence for a critical role for abdominal obesity is growing, the International Diabetes Federation (IDF) criteria [20] used an increased waist circumference as a prerequisite to replace insulin resistance. In recent years, clinical criteria have been largely harmonized [18]. This is reflected in the American Heart Association (AHA)/National Heart, Lung, and Blood Institute (NHLBI) update of the National Cholesterol Education Program (NCEP) criteria [16], and IDF recommendations [20]. The original NCEP threshold for elevated glucose was 110 mg/dL; at this cut point, only about 15% of the US population had a high glucose. In 2005, the AHA/NHLBI lowered the glucose threshold to 100 mg/dL [16]. This change led to an increase in elevated glucose to a level comparable to that of other risk factors. As a result of this change, the overall prevalence of the MetS was raised by about 6% [15]. Another concern is that a single cut point of central obesity is not suitable for all populations. Therefore, waist circumference threshold was modified for Asians as ≥90 cm for men and ≥80 cm for women in the AHA/NHLBI guidelines [16]. During the past decade, a large number of epidemiological studies have reported prevalence of the MetS for adults in different populations [15, 21–23]; the majority of studies have used NCEP Adult Treatment Panel III (ATP III) criteria [24]. However, standard and identical cut-off points are not available to define the MetS in the pediatric population [25]. In the BHS, selected percentiles of the MetS components are suggested, e.g. >75th percentiles (or <25th percentile for HDL cholesterol). In this chapter, we will focus on the statistical methodologies and findings of the MetS from the BHS.

4.3 Pathogenesis of MetS

Over the past three decades, considerable efforts have been made to understand pathogenesis of the MetS. Insulin resistance (i.e., resistance to insulin-stimulated glucose uptake) as a key feature of type 2 diabetes was first described many years earlier [26], and hyperinsulinemia was found to be associated with obesity, hypertension and dyslipidemia. Therefore, insulin resistance was recognized as the dominant cause of the clustering of these risk factors, and the condition was termed as "insulin resistance syndrome" [9]. In recent years, more and more evidence has been emerging for the role of obesity in the development of MetS [27] and morbidity and mortality from coronary heart disease [28]. Although the pathophysiologic mechanisms underlying the development of MetS are not fully understood, obesity is generally considered a primary defect underlying the clustering of the MetS components [27, 29–33]. In longitudinal cohorts of children, adolescents and young adults enrolled in the BHS with a follow-up period of 3 years, a significant and positive association was observed between the degree of baseline obesity and the incidence of hyperinsulinemia at follow-up in all three age groups which was independent of race, gender and baseline insulin levels. These results suggest that obesity may precede hyperinsulinemia beginning in childhood [29]. In another BHS longitudinal cohort followed from childhood to adulthood during a period of 11.6 years, childhood BMI was significantly associated with adulthood MetS even after adjusting for childhood insulin. In contrast, corresponding association with childhood insulin disappeared after adjusting for childhood BMI [30]. In a cross-sectional study of preadolescents, adolescents and young adults from the BHS, the degree of clustering of multiple MetS components was measured by the intra-class correlation coefficients; the adjustment for BMI resulted in ~50% reduction in the intra-class correlations in all three age groups; and the age-related pattern was no longer evident [32]. With growing evidence for a critical role for abdominal obesity, an increased waist circumference has gained more importance in the diagnostic criteria [20, 24]. Recently, the waist-to-height ratio is considered a better indicator of central obesity and more sensitive than BMI or waist circumference alone to evaluate clustering of the MetS variables [34]. In the BHS we also noted that the waist-to-height ratio was strongly associated with multiple MetS risk variables and related adverse C-V risk profile among normal weight younger adults [35].

4.4 Tracking of MetS

C-V risk variables change dramatically with age; however, longitudinal epidemiologic studies have demonstrated that the percentiles of the C-V risk variables in the population tend to remain relatively stable over a long period of time [3, 36–40]. The stability over time of an individual's risk variable levels relative to that of his/her peers is referred to as "tracking". The concept of "tracking" was established early in 1970s mainly based on the observations from the BHS and Muscatine Study [3, 36]. In longitudinal analyses of the BHS, the C-V risk variables were demonstrated to track not only during childhood [3, 37, 38] but also during a long period of follow-up from childhood to young adulthood [41–43].

In addition to the tracking of individual C-V risk factor variables over time, the clustering of multiple C-V risk variables was also noted to persist from childhood to young adulthood [44]. In a longitudinal cohort of black and white children who were followed for 8 years, a multiple risk index score was constructed by adding age-, sex- and race-specific rankings of three risk variables, systolic blood pressure, insulin and total-to-HDL cholesterol ratio, to measure the clustering of these variables. The multiple risk index was shown to track in all four race–sex groups (year 1 versus year 8 correlation: $r=0.54$–0.67). The overall multiple risk index tracking correlation ($r=0.64$) was significantly stronger than those for individual risk factors ($r=0.34$–0.57). Tracking of the multiple risk index increased progressively with age and adiposity.

Tracking correlations between baseline and follow-up values form the basis of the tracking phenomenon of individual variables. On the other hand, the tracking of clusters of multiple risk variables also results from the correlations of rates of change of these variables over time as evidenced by the observation from the BHS [45]. The clustering of long-term rates of change of the MetS components (body mass index, insulin resistance, triglycerides-to-HDL cholesterol ratio and mean arterial pressure) from childhood to adulthood was evaluated longitudinally (1982–2003) in a cohort of 1020 subjects who were examined 3–6 times both as children (ages 4–17 years) and as adults (ages 18–38 years) over an average of 16 years, with 3874 observations. The incremental area under the growth curve was used as a measure of long-term rates of change in risk variables since childhood (Fig. 4.1). Significant clustering (as measured by intra-class correlation) of the four variables was noted for childhood and adulthood values as well as the rates of change from childhood to adulthood. Adjustment for body mass index considerably reduced the degree of clustering of the rates of change (Fig. 4.2) [45]. In another study from

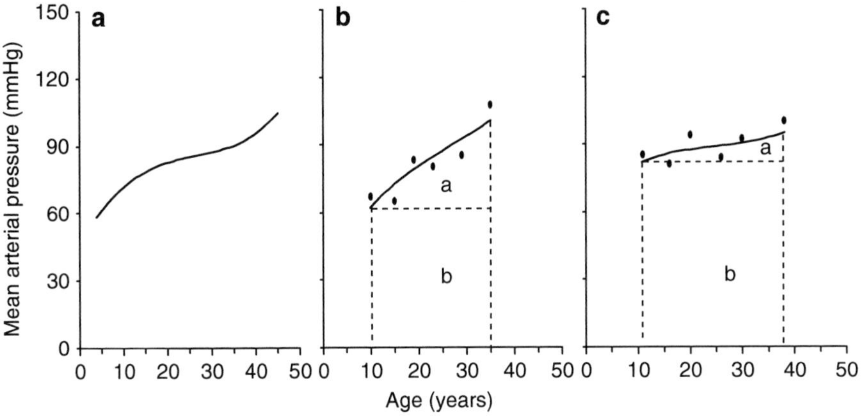

Fig. 4.1 Area under the curve of mean arterial pressure (MAP) as derived from a growth curve model, The Bogalusa Heart Study. **a** Overall growth curve (fixed parameters; MAP=45+0.54 age−0.25 age^2+0.08 age^3). **b** Curve for individual #1, examined at ages 10, 15, 19, 23, 29, and 35 years (random effects incorporated; MAP=31+1.4 age−0.23 age^2+0.06 age^3). **c** Curve for individual #2, examined at ages 11, 16, 20, 26, 30, and 38 years (random effects incorporated; MAP=66+0.3 age−0.11 age^2+0.05 age^3). *a,* incremental area; *b,* baseline area. [45]

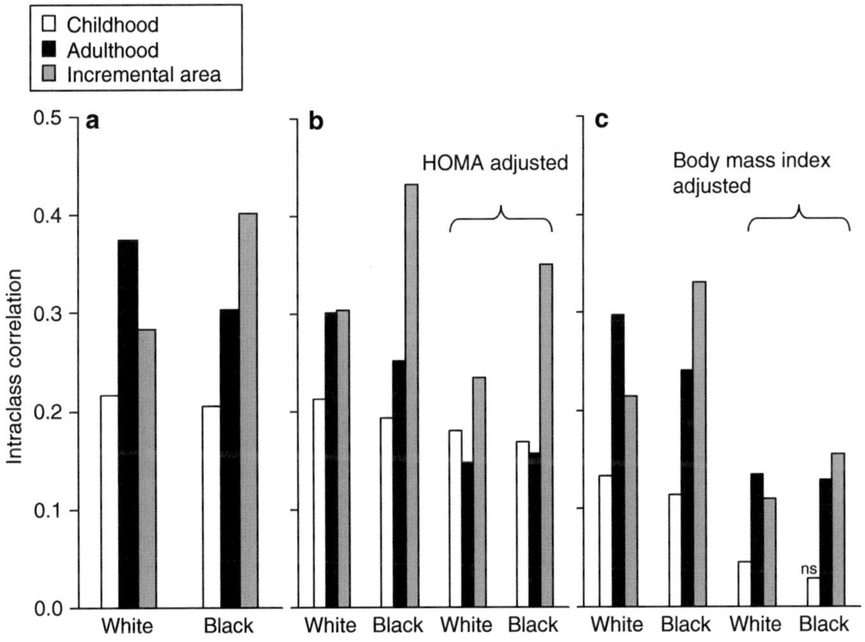

Fig. 4.2 Intraclass correlations of metabolic syndrome variables in terms of childhood, adulthood and incremental area values by race: The Bogalusa Heart Study. **a** Four variables. **b** Body mass index, mean arterial pressure and triglycerides/HDL cholesterol. **c** HOMA, mean arterial pressure and triglycerides/HDL cholesterol. *HOMA* homeostasis model assessment of insulin resistance; *ns* not significant ($p>0.05$). All intraclass correlation coefficients are significantly greater than zero ($p<0.05$) except for that marked ns, adjusting for age, sex and/or HOMA or body mass index. [45]

the BHS, Srinivasan et al also reported that the rate of change in adiposity was associated with changes in C-V risk variables over 6–8 years during both childhood and adulthood [31]. Further, the MetS components were found to change together over time in a 4.5-year follow-up study during adulthood [33]. These observations demonstrated that the MetS variables coexist, in terms not only of their levels in childhood and adulthood but also of long-term rates of change. Obesity is of critical importance in the development of the MetS, and its prevention beginning in childhood needs to be addressed.

4.5 Clustering Analysis

4.5.1 Observed to Expected Ratio

The high prevalence of clusters of multiple risk variables should not be considered evidence of clustering because the high frequencies of individual component disorders would result in more subjects with multiple disorders by chance alone. The phenomenon of clustering of risk variables is defined as multiple risk variables that are present in the same individual more often than by chance alone as one would expect. The commonly used approach to assess the clustering of MetS components beyond chance is observed to expected (O/E) ratio [46–48]. The quantitative MetS components have to be transformed into dichotomous disorders using cut-off values. Under the assumption of independence of the components, the expected is calculated by multiplying the prevalence of each disorder included in the cluster and term(s) of 1 minus the prevalence of the disorder(s) not included in the cluster. The observed is the actual prevalence of corresponding cluster. The O/E ratio is evidence of clustering because it measures how many times the clustering of MetS components goes beyond chance. Significance tests for O/E ratios can be performed using generalized one-sample binomial test when $E \geq 5$ and using the Poisson distribution when $E < 5$ [49].

For example, the prevalence rates of hyperinsulinemia, obesity, high blood pressure and dyslipidemia were 26.4%, 23.9%, 25.0% and 24.4%, respectively. The expected frequency of subjects having hyperinsulinemia, obesity and high blood pressure without dyslipidemia was computed as $0.264 \times 0.239 \times 0.250 \times (1-0.244) \times 100 = 1.19\%$. The frequency 1.19% does not include the cluster of four disorders. In compliance with the calculation of the expected frequency, the observed frequency of the cluster had to be defined as the percent of coexistence of these three isolated disorders without dyslipidemia to make all clusters mutually exclusive. The observed frequency of the cluster hyperinsulinemia, obesity and high blood pressure without dyslipidemia was 3.12%; a statistically significant excess of this cluster was indicated by an O/E ratio of 2.61 ($P < 0.05$) [46].

The O/E ratio approach was used to assess the clustering of three and four disorders in parents and their young offspring enrolled in the BHS [46]. The disorders were defined by greater than 75th age-, sex- and race-specific percentiles of fasting insulin, body mass index, triglycerides-to-HDL cholesterol ratio and mean arterial pressure. Based on O/E ratios, there was a significant excess of offspring

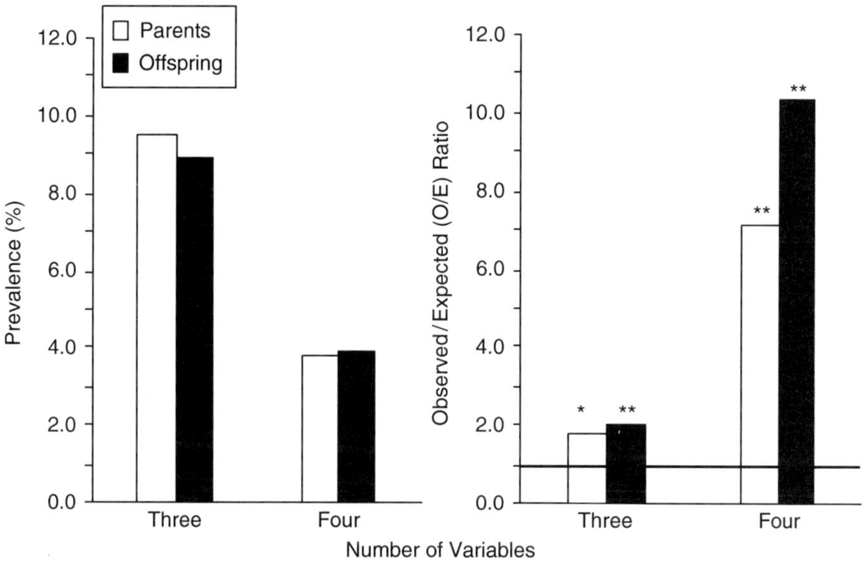

Fig. 4.3 Prevalence and observed/expected (O/E) ratio of clustering of three and four disorders in parents and their young offspring. Disorders were defined by greater than 75th age-, sex- and race-specific percentiles of fasting insulin, body mass index, triglycerides-to-HDL cholesterol ratio and mean arterial pressure. O/E ratios greater than 1 under the null hypothesis: *$p<0.05$; **$p<0.01$. [46]

and parents with clusters of three and four disorders (Fig. 4.3). Of note, although the prevalence of three disorder cluster was substantially higher than that of four disorder cluster as one would expect, the O/E ratios were lower for the former than for the latter. The comparison of these two parameters provides an explanation of why O/E ratio rather than prevalence has to be used as a measure of clustering. A higher prevalence of the clustering cannot be considered evidence for existence of the MetS components. Significant O/E ratios of clustering of the above four MetS components were also observed in all age groups of children and adults from the BHS [32].

4.5.2 Correlation Patterns

The O/E ratio is easy to calculate and commonly used to estimate the degree of the clustering of risk variables. However, this measure has limitations and is influenced by the diagnostic or percentile cut-off values chosen to define the abnormal risk variables. Therefore, it is difficult to compare the O/E ratios among studies using different cut-offs to define the MetS. Intra-class correlation (ICC) of multiple continuous variables can overcome the weakness of using cut points for the MetS risk variables. We introduced the ICC approach to evaluate the degree of clustering of two or more continuous risk variables in the BHS population [32]. The ICC is defined as the proportion of between-subject variance in the total variance, a sum of between-subject variance and within-subject variance which are estimated using between-subject mean square (BMS) and within-subject mean square (WMS), respectively, of the sample. In one-way analysis of variance, the ICC coefficient can be calculated by the expression [32, 50]:

$$ICC = (BMS - WMS) / [BMS + (k - 1) WMS],$$

where k is the number of risk variables.

The ICC coefficient measures how closely subjects maintain their rankings of multiple risk variables by comparing within-subject and between-subject variations in the MetS variables. With higher ICC, the within-subject variability is smaller compared to the between-subject variability. Therefore, the higher the intra-class correlation, the stronger the clustering of risk variables. Since, unlike O/E ratios, the ICC calculation involves continuous variables that obviate problems associated with the definition of cut-off points, we used this approach to examine the correlation patterns of various combinations of the MetS risk variables by age groups ranging 5–37 years in a cross-sectional analysis of 7875 observations from the BHS. Among the 6 pairs of the MetS components, the highest correlation was seen for BMI and insulin resistance, and lowest for blood pressure and triglycerides-to-HDL cholesterol ratio in all age groups. For all four variables, the ICC ranged from 0.142 to 0.399 ($p<0.001$), with adolescents showing the lowest ICC. When BMI was adjusted for, the ICC involving the other three variables was reduced by about 50%, and the age-related pattern was no longer evident (Fig. 4.4) [32].

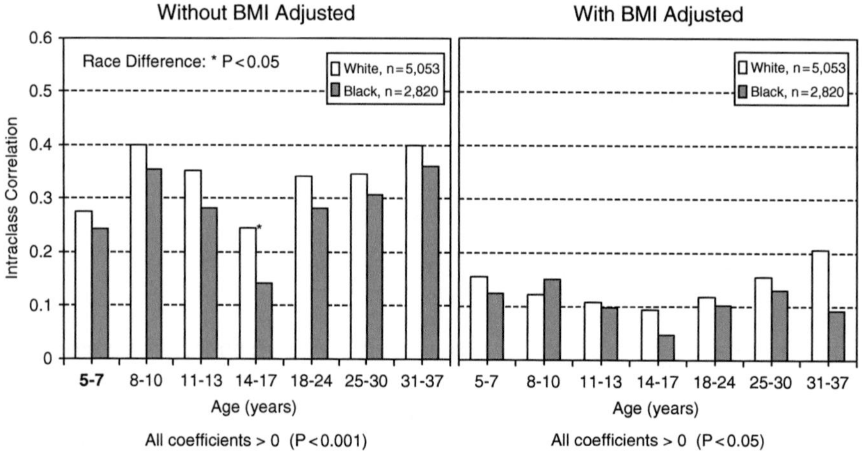

Fig. 4.4 Intra-class Correlations of BMI, insulin resistance index, mean arterial pressure and triglycerides-to-HDL cholesterol ratio by race and age: The Bogalusa Heart Study. *HOMA* homeostasis model assessment of insulin resistance. [32]

4.5.3 Clustering Features

The evidence for coexistence of multiple C-V risk variables is indisputable. However, the correlation structure of the MetS variables varies by different combinations of the variables, age groups and populations. The strong correlation between variables, like BMI and insulin resistance as noted above, suggests a common underlying factor. The technique of factor analysis, which resolves the highly interrelated variables into a set of composite factors unrelated to each other [51], is particularly useful for examining the multifaceted nature of the MetS and identifying the underlying pathophysiologic factors. Studies have shown 1–5 independent pathophysiologic process underlying risk factor clustering in pediatric and adult populations [25, 52–56]. In the BHS, factor analysis was applied to examine the clustering characteristics of the MetS in 4522 subjects including children (5–11 years), adolescents (12–17 years) and young adults (18–38 years). Adiposity, insulin, glucose, triglycerides, HDL cholesterol, systolic and diastolic blood pressures were used as components of the MetS. Factor analysis yielded two uncorrelated factors (Factor 1: insulin/lipids/glucose/ponderal index; Factor 2: insulin/blood pressure) (Fig. 4.5). These two factors explained 54.6% of the total variance in the entire sample. The factor loading patterns were very similar in all race, sex and age groups. These results suggest that the MetS is characterized by the linking of a metabolic entity (insulin resistance, dyslipidemia and obesity) to a hemodynamic factor (hypertension) through shared correlation with insulin resistance, and the clustering features are independent of sex and age in both black and white populations [56].

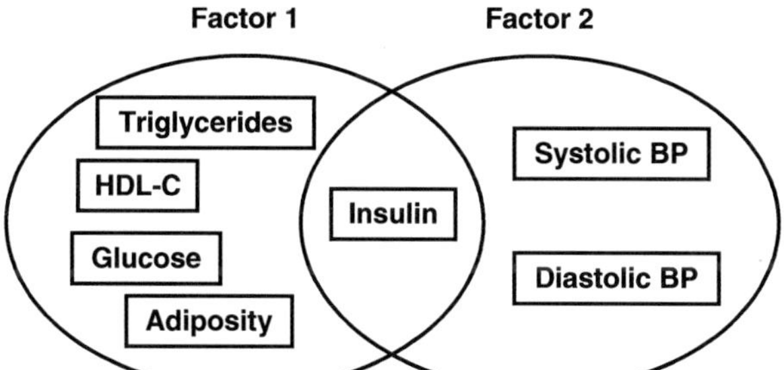

Fig. 4.5 Graphic interpretation of factor loading pattern of the MetS risk variables based on factor analysis: The Bogalusa Heart Study. The *small boxes* represent seven risk variables included in the analysis; the *large circles* represent two factors characterizing two distinct features of the MetS. The two features are linked by the shared correlations with hyperinsulinemia. Variables loaded on each factor were interpreted with loadings greater than or equal to 0.3. *HDL-C* high-density lipoprotein cholesterol; *BP* blood pressure. [56]

4.5.4 Complex Relationships

The MetS risk variables are highly inter-correlated with each other. The O/E ratio and intra-class correlation analyses discussed above provide evidence for overall clustering of multiple risk variables without information on detailed correlation structure among the variables. The complex interrelationships of multiple MetS

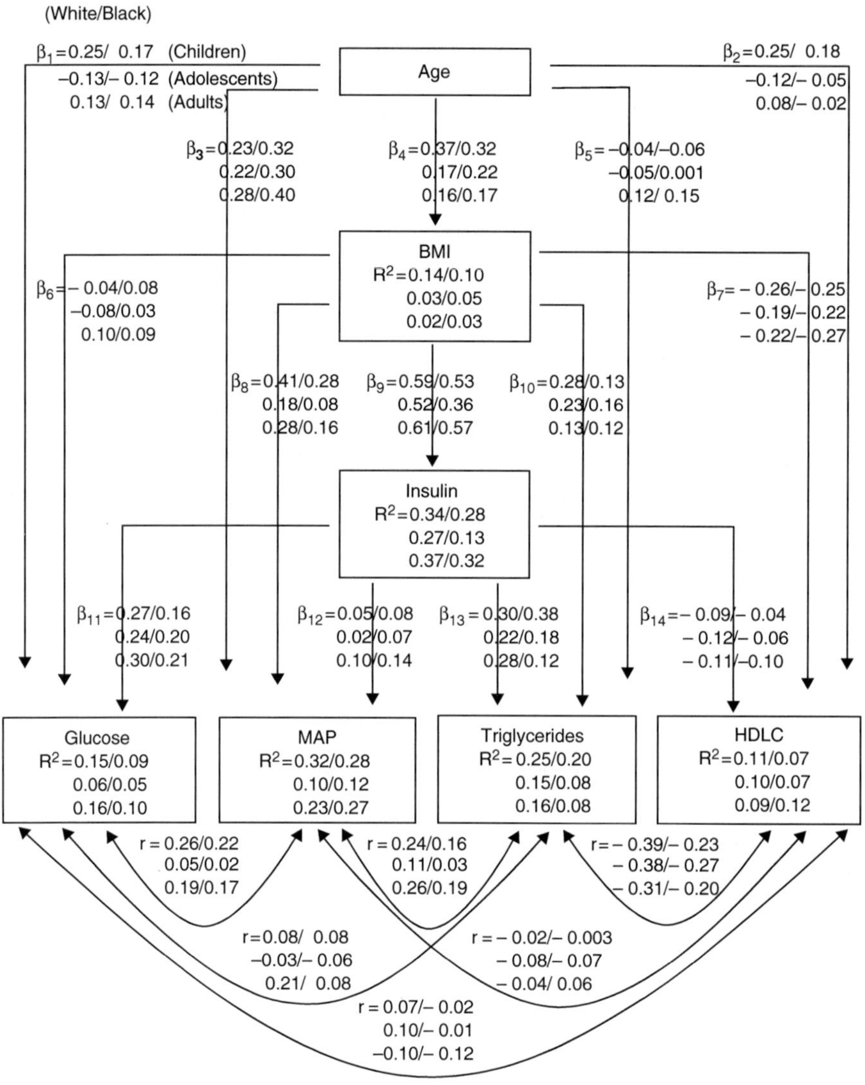

Fig. 4.6. Relationships among metabolic syndrome components in white and black children, adolescents and adults: The Bogalusa Heart Study. *BMI* body mass index; *MAP* mean arterial pressure; *HDL-C* high-density lipoprotein cholesterol; β_i path coefficient; R^2 variance explained; *r* Pearson correlation coefficients. Correlation coefficients are significant ($p<0.05$) when $|r|>0.045$ for whites and $|r|>0.058$ for blacks. [60]

Table 4.1 Selected path coefficients in children, adolescents and adults by race. [60]

	Children			Adolescents			Adults		
	White	Black	P†	White	Black	P†	White	Black	P†
β_3	0.23**	0.32**	0.294	0.22**	0.30**	0.308	0.28**	0.40**	*0.010*
β_6	−0.04	0.08*	0.075	−0.08**	0.03	*0.048*	0.10**	0.09*	1.000
β_8	0.41**	0.28**	*0.007*	0.18**	0.08**	0.090	0.28**	0.16**	*0.022*
β_9	0.59**	0.53**	0.520	0.52**	0.36**	*<0.001*	0.61**	0.57**	1.000
β_{10}	0.28**	0.13**	*0.003*	0.23**	0.16**	0.742	0.13**	0.12**	1.000
β_{13}	0.30**	0.38**	0.686	0.22**	0.18**	1.000	0.28**	0.12**	*<0.001*

Path coefficients are different from zero: *p<0.05; **p<0.01
† Bonferroni adjusted p value for race difference

components cannot be elucidated without using an advanced sophisticated statistical method, like path analysis.

Path analysis is a multivariate analysis capable of dissecting the causal relationships among observed inter-correlated variables. The various aspects of formulating, fitting and testing such relationships are also referred to as "linear structural equation modeling". Path analysis has become a prominent form of data analysis in the fields of sociology, economics, ecology and psychology [57, 58]. Software packages of EQS and LISREL and procedures in SAS, SPSS and BMDP have been developed specifically for linear structural equation modeling analysis. In addition to the extensive use in family data for genetic analyses [59], this approach has been previously used to examine the interrelationships among the MetS components [60–63]. Path analysis has advantages over the conventional multivariate regression analysis in that (1) both direct effects and indirect effects through other variables or confounders can be estimated, and (2) unlike separate regression analyses, path analysis can estimate multiple parameters simultaneously in a multi-stage model. Furthermore, the significance of difference in multiple path coefficient parameter estimates between the two models can be tested [57, 64, 65]. Path analysis promises to be very useful in the research area of MetS.

Figure 4.6 shows the complex relationships among age, BMI, insulin, glucose, triglycerides, HDL cholesterol, mean arterial pressures by race and age groups using path analysis in 8203 black and white individuals enrolled in the BHS [60]. In general, path coefficients were greater in whites than in blacks except for the age-mean arterial pressure path (Table 4.1), and greater in children and adults than in adolescents. The black–white differences in the relationships of obesity and insulin resistance measures to other components may account for the lower prevalence of the MetS in the black population.

4.6 Black–white Difference in Prevalence of MetS

Although blacks have higher prevalence rates of type 2 diabetes [66, 67], studies in both children and adults showed lower prevalence of MetS in blacks [68, 69]. The ARIC study reported that African–American subjects had significantly lower OR than white subjects for the association of diabetes with clustering of the remaining components [48]. It is proposed that lower triglycerides and higher HDL cholesterol levels in blacks, and lower waist circumference in black men lead to underdiagnosis of MetS in blacks [69].

The prevalence of the MetS depends on two determinants, i.e. the frequency of individual component disorders (defined by the cut points) and the magnitude of correlation among the components. For example, frequencies of three disorders are all 0.1 in population A. Under the assumption of no clustering, the expected prevalence of MetS is $0.1^3=0.1\%$ in this population. In population B, frequencies of the 3 disorders are all 0.2. The expected prevalence of MetS is $0.2^3=0.8\%$. In the case of strong correlation among the risk variables, the observed prevalence is supposed to be higher than the expected. The significantly lower triglycerides and higher HDL cholesterol definitely contribute in part to the lower prevalence of MetS in the black population. However, the fact that black adults also have higher blood pressure and insulin resistance [32, 45, 46, 70] points to the importance of the strength of the correlation between triglycerides and HDL cholesterol and their correlations with other risk variables in this regard. We found that the degree of clustering of the MetS components measured as O/E ratio and ICC (Fig. 4.4) was lower in blacks than in whites almost in all age groups, particularly in adolescents [32]. In addition, majority of the pair-wise correlations (Table 4.2) and path coefficients (Fig. 4.6) were lower in blacks than in whites [56, 60]. Clearly, data from the BHS provide evidence that the lower prevalence of MetS in the black population may largely result from the lower correlations among the MetS variables.

4.7 MetS and C-V Risk

The MetS has gained increasing importance because of its association with subsequent morbidity and mortality from C-V disease. The risk for C-V disease accompanying the MetS is approximately doubled [16]. For example, a recent meta-analysis including 43 cohorts (172,573 individuals) reported that the MetS conveyed a relative risk for C-V events and death of 1.78 [71]. The NCEP in its most recent ATP III report recognized MetS as a highly important C-V risk entity [11]. In the BHS, the MetS was found to be associated with C-V risk measured by subclinical changes of the C-V system [72–74]. Increased left ventricular mass indexed to height ($m^{2.7}$) and relative wall thickness were associated with the MetS [72]. In a study cohort of 1073 black and white subjects aged 25 to 44 years, individuals who had 3 or 4 MetS components had significantly higher carotid artery intima-media thickness (IMT) in both blacks and whites; the effect of MetS on carotid IMT was modified by the parental history of coronary artery disease, with a significant greater effect when the parental history was positive [73]. Further, clustering of the MetS variables at low levels in childhood were found to be beneficially associated with adulthood C-V risk measured by carotid artery IMT in the BHS [70]. Strong influence of the MetS on arterial stiffness measured as pulse wave velocity was also noted in the BHS cohort [74]. In addition, strong associations of the MetS with chronic systemic

Table 4.2 Age- and sex-adjusted pearson correlations coefficients among MetS risk variables by race and age group. [56]

	Diastolic BP		Log-TG		HDL-C		Glucose		Log-insulin		Adiposity	
	Black	White	Black	White	Black	White	Black	White	Black	White	Black	White
Children												
Systolic BP	0.54	0.61	0.16	0.32	–	−0.12	0.17	0.33	0.32	0.42	0.28	0.43
Diastolic BP			–	0.24	–	−0.12	–	0.21	0.21	0.29	0.20	0.34
Log-TG					−0.22	−0.41	–	0.17	0.42	0.43	0.29	0.39
HDL-C							–	–	−0.17	−0.23	−0.22	−0.28
Glucose									0.33	0.35	–	0.16
Log-Insulin											0.44	0.57
Adolescents												
Systolic BP	0.41	0.52	–	0.14	–	–	–	0.15	0.20	0.25	0.12	0.20
Diastolic BP			–	0.10	–	–	–	0.12	–	0.20	–	0.20
Log-TG					−0.28	−0.42	–	–	0.33	0.44	0.25	0.40
HDL-C							−0.13	–	−0.26	−0.30	−0.33	−0.30
Glucose									0.19	0.34	0.13	0.10
Log-Insulin											0.45	0.61
Adults												
Systolic BP	0.66	0.70	0.12	0.25	–	–	0.11	0.09	0.22	0.28	0.24	0.29
Diastolic BP			0.12	0.24	–	−0.07	–	0.11	0.21	0.27	0.23	0.30
Log-TG					−0.26	−0.38	–	0.14	0.36	0.49	0.23	0.39
HDL-C							−0.16	−0.09	−0.36	−0.29	−0.31	−0.28
Glucose									0.38	0.27	0.26	0.19
Log-Insulin											0.59	0.63

– Correlation coefficients with $p > 0.01$, *BP* blood Pressure, *HDL-C* high-density lipoprotein cholesterol, *TG* triglycerides inflammation status, a novel risk factor of C-V disease, reinforce the importance of the MetS in assessing the C-V risk. Data from the BHS showed that the MetS was significantly associated with C-reactive protein, white blood cells and adiponectin [68, 75, 76]. Given the consistency of data for C-reactive protein, researchers have suggested that C-reactive protein should be added to MetS for assessment of global C-V risk [77].

4.8 Conclusions

The prevalence of the MetS is a substantial public health burden worldwide and has an increasing trend. According to NCEP ATP III definition, about 47 million US residents had MetS in 1990 and 64 million had the syndrome in 2000 [78, 79]. This increase parallels rise in the prevalence of obesity. The likelihood of a further increase in the MetS can be anticipated because of projections of a greater prevalence of obesity in the future [80]. We also found that in semirural Bogalusa, the childhood obesity epidemic has not plateaued, and nearly half of the children are now overweight or obese [81]. Importantly, multiple risk factors interact with each other and persist (track) from childhood into adulthood. The MetS may even originate in the embryonic and fetal stages [82, 83]. Observations from the BHS reinforce recommendations to prevent and modulate the development of the MetS during childhood. Controlling weight gain of children through improved nutrition and physical activity becomes the main direction for prevention. Importantly, the BHS provides the evidence of the urgency of achieving a public health policy to begin the prevention of adult MetS in childhood.

References

1. Haller H (1977) Epidemiology and associated risk factors of hyperlipoproteinemia. Z Gesamte Inn Med 32:124–128
2. Webber LS, Voors AW, Srinivasan SR, Frerichs RR, Berenson GS (1979) Occurrence in children of multiple risk factors for coronary artery disease: the Bogalusa Heart Study. Prev Med 8:407–418
3. Berenson GS, Webber LS, Srinivasan SR, Frerichs RR, Voors AW (1978) Interrelationship and persistence of risk factor variables at high levels in children—the Bogalusa Heart Study. In: Carlson LA, Paoletti R, Sirtori CR, Weber G (eds) International conference on atherosclerosis. Raven Press, New York, pp 357–363
4. Smoak CG, Burke GL, Harsha DW, Srinivasan SR, Berenson GS (1987) Relation of obesity to clustering of cardiovascular disease risk factors in children and young adults: the Bogalusa Heart Study. Am J Epidemiol 125:364–372
5. McGill HC Jr, McMahan CA, Zieske AW, Tracy RE, Malcom GT, Herderick EE, Strong JP (2000) Association of coronary heart disease risk factors with microscopic qualities of coronary atherosclerosis in youth. Circulation 102:374–379

6. Berenson GS, Srinivasan SR, Bao W, Newman WP III, Tracy RE, Wattigney WA (1998) Association between multiple cardiovascular risk factors and atherosclerosis in children and young adults. New Eng J Med 338:1650–1656
7. Berenson GS, McMahan CA, Voors AW (1980) Cardiovascular risk factors in children. The early natural history of atherosclerosis and essential hypertension. Oxford University Press, Oxford
8. Reaven GM (1988) Role of insulin resistance in human disease. Diabetes 37:1595–1607
9. Haffner SM, Valdez RA, Hazuda HP et al (1992) Prospective analysis of the insulin-resistance syndrome (syndrome X). Diabetes 41:715–722
10. World Health Organization (1999) Definition, diagnosis, and classification of diabetes mellitus and its complications: report of a WHO consultation. Geneva, World Health Organization
11. Expert Panel on the Detection, Evaluation, and Treatment of High Blood Cholesterol in Adults (2001) Executive summary of the Third Report of the National Cholesterol Education Program (NCEP) Expert Panel on detection, evaluation, and treatment of high blood cholesterol in adults (Adult Treatment Panel III). JAMA 285:2486–2497
12. Grundy SM, Brewer HB Jr, Cleeman JI, Smith SC Jr, Lenfant C, National Heart, Lung, and Blood Institute, American Heart Association (2004) Definition of metabolic syndrome: report of the National Heart, Lung, and Blood Institute/American Heart Association conference on scientific issues related to definition. Circulation 109:433–438
13. Kahn R, Buse J, Ferrannini E, Stern M (2005) The metabolic syndrome: time for a critical appraisal: joint statement from the American Diabetes Association and the European Association for the Study of Diabetes. Diabetes Care 28:2289–2304
14. Cameron AJ, Shaw JE, Zimmet PZ (2004) The metabolic syndrome: prevalence in worldwide populations. Endocrinol Metab Clin North Am 33:351–375
15. Grundy SM (2008) Metabolic syndrome pandemic. Arterioscler Thromb Vasc Biol 28:629–636
16. Grundy SM, Cleeman JI, Daniels SR, Donato KA, Eckel RH, Franklin BA, Gordon DJ, Krauss RM, Savage PJ, Smith SC Jr, Spertus JA, Costa F, American Heart Association, National Heart, Lung, and Blood Institute (2005) Diagnosis and management of the metabolic syndrome: an American Heart Association/National Heart, Lung, and Blood Institute scientific statement. Circulation 112:2735–2752
17. Steinberger J, Daniels SR, Eckel RH, Hayman L, Lustig RH, McCrindle B, Mietus-Snyder ML, American Heart Association Atherosclerosis, Hypertension, and Obesity in the Young Committee of the Council on Cardiovascular Disease in the Young, Council on Cardiovascular Nursing, Council on Nutrition, Physical Activity, and Metabolism (2009) Progress and challenges in metabolic syndrome in children and adolescents: a scientific statement from the American Heart Association Atherosclerosis, Hypertension, and Obesity in the Young Committee of the Council on Cardiovascular Disease in the Young; Council on Cardiovascular Nursing; and Council on Nutrition, Physical Activity, and Metabolism. Circulation 119:628–647
18. Alberti KG, Eckel RH, Grundy SM, Zimmet PZ, Cleeman JI, Donato KA, Fruchart JC, James WP, Loria CM, Smith SC Jr, International Diabetes Federation Task Force on Epidemiology and Prevention, Hational Heart, Lung, and Blood Institute, American Heart Association, World Heart Federation, International Atherosclerosis Society, International Association for the Study of Obesity (2009) Harmonizing the metabolic syndrome: a joint interim statement of the International Diabetes Federation Task Force on Epidemiology and Prevention; National Heart, Lung, and Blood Institute; American Heart Association; World Heart Federation; International Atherosclerosis Society; and International Association for the Study of Obesity. Circulation 120:1640–1645
19. Alberti KG, Zimmet PZ (1998) Definition, diagnosis and classification of diabetes mellitus and its complications. Part 1: diagnosis and classification of diabetes mellitus provisional report of a WHO consultation. Diabetes Med 15:539–553
20. Alberti KG, Zimmet P, Shaw J, IDF Epidemiology Task Force Consensus Group (2005) The metabolic syndrome—a new worldwide definition. Lancet 366:1059–1062

21. Feng Y, Hong X, Li Z, Zhang W, Jin D, Liu X, Zhang Y, Hu FB, Wei LJ, Zang T, Xu X, Xu X (2006) Prevalence of metabolic syndrome and its relation to body composition in a Chinese rural population. Obesity (Silver Spring) 14:2089–2098
22. Thomas GN, Ho SY, Janus ED, Lam KS, Hedley AJ, Lam TH (2005) The US National Cholesterol Education Program Adult Treatment Panel III (NCEP ATP III) prevalence of the metabolic syndrome in a Chinese population. Diabetes Res Clin Pract 67:251–257
23. Park YW, Zhu S, Palaniappan L, Heshka S, Carnethon MR, Heymsfield SB (2003) The MetS: prevalence and associated risk factor findings in the US population from the Third National Health and Nutrition Examination Survey, 1988–1994. Arch Intern Med 163:427–436
24. Third report of the National Cholesterol Education Program (NCEP) expert panel on detection, evaluation, and treatment of high blood cholesterol in adults (Adult Treatment Panel III). Final Report. Circulation 2002 106:3143–3421
25. Ford ES, Li C (2008) Defining the metabolic syndrome in children and adolescents: will the real definition please stand up? J Pediatr 152:160–164
26. Himsworth H (1936) Diabetes mellitus: a differentiation into insulin-sensitive and insulin-insensitive types. Lancet 1:127–130
27. Despres JP, Lemieux I (2006) Abdominal obesity and metabolic syndrome. Nature 444:881–887
28. Must A, Jacques PF, Dallal GE, Bajema CJ, Dietz WH (1992) Long-term morbidity and mortality of overweight adolescents. A follow-up of the Harvard Growth Study of 1922 to 1935. N Engl J Med 327:1350–1355
29. Srinivasan SR, Myers L, Berenson GS (1999) Temporal association between obesity and hyperinsulinemia in children, adolescents, and young adults: the Bogalusa Heart Study. Metabolism 48:928–934
30. Srinivasan SR, Myers L, Berenson GS (2002) Predictability of childhood adiposity and insulin for developing insulin resistance syndrome (syndrome X) in young adulthood: the Bogalusa Heart Study. Diabetes 51:204–209
31. Srinivasan SR, Myers L, Berenson GS (2001) Rate of change in adiposity and its relationship to concomitant changes in cardiovascular risk variables among biracial (black–white) children and young adults: the Bogalusa Heart Study. Metabolism 50:299–305
32. Chen W, Bao W, Begum S, Elkasabany A, Srinivasan SR, Berenson GS (2000) Age-related patterns of the clustering of cardiovascular risk variables of syndrome X from childhood to young adulthood in a population made up of black and white subjects: the Bogalusa Heart Study. Diabetes 49:1042–1048
33. Maison P, Byrne CD, Hales CN et al (2001) Do different dimensions of the metabolic syndrome change together over time? Evidence supporting obesity as the central feature. Diabetes Care 24:1758–1763
34. Hsieh SD, Muto T (2005) The superiority of waist-to-height ratio as an anthropometric index to evaluate clustering of coronary risk factors among non-obese men and women. Prev Med 40:216–220
35. Srinivasan SR, Wang R, Chen W, Wei CY, Xu J, Berenson GS (2009) Utility of waist-to-height ratio in detecting central obesity and related adverse cardiovascular risk profile among normal weight younger adults (from the Bogalusa Heart Study). Am J Cardiol 104:721–724
36. Clarke WR, Schrott HG, Leaverton PE, Connor WE, Lauer RM (1978) Tracking of blood lipids and blood pressures in school age children: the Muscatine Study. Circulation 58:626–634
37. Webber LS, Cresanta JL, Voors AW, Berenson GS (1983) Tracking of cardiovascular disease risk factor variables in school-age children. J Chronic Dis 36:647–660
38. Webber LS, Srinivasan SR, Wattigney WA, Berenson GS (1991) Tracking of serum lipids and lipoproteins from childhood to adulthood: the Bogalusa Heart Study. Am J Epidemiol 133:884–899
39. Porkka KVK, Viikari JSA, Taimela S, Dahl M, Åkerblom HK (1994) Tracking and predictiveness of serum lipid and lipoprotein measurements in childhood: a 12-year follow-up. The cardiovascular risk in Young Finns Study. Am J Epidemiol 140:1096–1110
40. Lane DA, Gill P (2004) Ethnicity and tracking blood pressure in children. J Hum Hypertens 18:223–228

41. Bao W, Threefoot S, Srinivasan SR, Berenson GS (1995) Essential hypertension predicted by tracking of elevated blood pressure from childhood to adulthood—the Bogalusa Heart Study. Am J Hyperten 8:657–665
42. Bao W, Srinivasan SR, Berenson GS (1996) Persistent elevation in insulin is associated with increased cardiovascular risk factors in children and young adults: the Bogalusa Heart Study. Circulation 93:54–59
43. Freedman DS, Khan LK, Serdula MK, Dietz WH, Srinivasan SR, Berenson GS (2005) Racial differences in the tracking of childhood BMI to adulthood. Obes Res 13:928–935
44. Bao W, Srinivasan SR, Wattigney WA, Berenson GS (1994) Persistence of multiple cardiovascular risk clustering related to syndrome X from childhood to young adulthood: the Bogalusa Heart Study. Arch Intern Med 154:1842–1847
45. Chen W, Srinivasan SR, Li S, Xu J, Berenson GS (2007) Clustering of long-term trends in metabolic syndrome variables from childhood to adulthood in Blacks and Whites: the Bogalusa Heart Study. Am J Epidemiol 166:527–533
46. Chen W, Srinivasan SR, Elkasabany A, Berenson GS (1999) The association of cardiovascular risk factor clustering related to insulin resistance syndrome (syndrome X) between young parents and their offspring: the Bogalusa Heart Study. Atherosclerosis 145:197–205
47. Liese AD, Mayer-Davis EJ, Tyroler HA, Davis CE, Keil U, Schmidt MI, Brancati FL, Heiss G (1997) Familial components of the multiple metabolic syndrome: the ARIC Study. Diabetologia 40:963
48. Schmidt MI, Duncan BB, Watson RL, Sharrett AR, Brancati FL, Heiss G (1996) A metabolic syndrome in whites and African–Americans. The Atherosclerosis risk in communities baseline study. Diabetes Care 19:414–418
49. Rosner B (2000) Fundamentals of biostatistics, 5th edn. Duxbury Press, Boston 249–260
50. Shrout PE, Fleiss JL (1979) Intra-class correlation: uses in assessing rater reliability. Psychol Bull 86:420–428
51. Harman HH (1976) Modern factor analysis. University of Chicago Press, Chicago
52. Meigs JB, D'Agostino RB Sr, Wilson PWF et al (1997) Risk variables clustering in the insulin resistance syndrome: the Framingham Offspring Study. Diabetes 46:1594–1600
53. Edwards KL, Burchfiel CM, Sharp DS et al (1998) Factors of the insulin resistance syndrome in nondiabetic elderly Japanese–American men. Am J Epidemiol 147:441–447
54. Donahue RP, Bean JA, DcCarlo Donahue R et al (1997) Does insulin resistance unite the separate components of the insulin resistance syndrome? Evidence from the Miami Community Health Study. Arterioscler Thromb Vasc Biol 17:2413–2417
55. Gray RS, Fabsitz RR, Cowan LD et al (1998) Risk factor clustering in the insulin resistance syndrome: the Strong Heart Study. Am J Epidemiol 148:869–878
56. Chen W, Srinivasan SR, Elkasabany A, Berenson GS (1999) Cardiovascular risk factors clustering features of insulin resistance syndrome (syndrome X) in a biracial (Black–White) population of children, adolescents, and young adults: the Bogalusa Heart Study. Am J Epidemiol 150:667–674
57. Bollen KA, Long LS (1993) Testing structural equation models. SAGE Publications Inc., Newbury Park
58. Li CC (1975) Path analysis—a primer. Boxwood Press, Pacific Grove
59. Rao DC, Morton NE, Yee S (1976) Resolution of cultural and biological inheritance by path analysis. Am J Hum Genet 28:228–242
60. Chen W, Srinivasan SR, Berenson GS (2008) Path analysis of metabolic syndrome components in black versus white children, adolescents and adults: the Bogalusa Heart Study. Ann Epidemiol 18:85–91
61. Chan JC, Cheung JC, Lau EM (1996) The metabolic syndrome in Hong Kong Chinese. The interrelationships among its components analyzed by structural equation modeling. Diabetes Care 19:953–959
62. Vitaliano PP, Scanlan JM, Zhang J, Savage MV, Hirsch IB, Siegler IC (2002) A path model of chronic stress, the metabolic syndrome, and coronary heart disease. Psychosom Med 64:418–435

63. Niaura R, Banks SM, Ward KD et al (2000) Hostility and the metabolic syndrome in older males: the normative aging study. Psychosom Med 62:7–16
64. Joreskog K, Sorbom D (1993) LISREL 8.52: structural equation modeling with the SIMPLIS command language. Scientific Software International, Chicago
65. Joreskog K, Sorbom D (2001) LISREL 8.52: user's reference guide. Scientific Software International, Chicago
66. Egede LE, Dagogo-Jack S (2005) Epidemiology of type 2 diabetes: focus on ethnic minorities. Med Clin North Am 89:949–975
67. Carter JS, Pugh JA, Monterrosa A (1996) Non-insulin-dependent diabetes mellitus in minorities in the United States. Ann Intern Med 125:221–232
68. Chen W, Srinivasan SR, Xu J, Berenson GS (2010) Black–white divergence in the relation of white blood cell count to aspects of metabolic syndrome in preadolescents, adolescents and young adults: the Bogalusa Heart Study. Diabetes Care 31:2474–2476
69. Sumner AE (2009) Ethnic differences in triglyceride levels and high-density lipoprotein lead to underdiagnosis of the metabolic syndrome in black children and adults. J Pediatr 155(Suppl 7):e7–e11
70. Chen W, Srinivasan SR, Li S, Xu J, Berenson GS (2005) Metabolic syndrome variables at low levels in childhood are beneficially associated with adulthood cardiovascular risk: the Bogalusa Heart Study. Diabetes Care 28:126–131
71. Gami AS, Witt BJ, Howard DE, Erwin PJ, Gami LA, Somers VK, Montori VM (2007) Metabolic syndrome and risk of incident cardiovascular events and death: a systematic review and meta-analysis of longitudinal studies. J Am Coll Cardiol 49:403–414
72. Patel DA, Srinivasan SR, Chen W, Berenson GS (2009) Influence of the metabolic syndrome versus the sum of its individual components on left ventricular geometry in young adults (from the Bogalusa Heart Study). Am J Cardiol 104:69–73
73. Chen W, Srinivasan SR, Xu J, Berenson GS (2008) Effect of parental coronary artery disease on adverse effects of the metabolic syndrome and aging on carotid artery intima-media thickness (from the Bogalusa Heart Study). Am J Cardiol 102:180–183
74. Li S, Chen W, Srinivasan SR, Berenson GS (2005) Influence of metabolic syndrome on arterial stiffness and its age-related change in young adults: the Bogalusa Heart Study. Atherosclerosis 180:349–354
75. Patel DA, Srinivasan SR, Xu JH, Li S, Chen W, Berenson GS (2006) Distribution and metabolic syndrome correlates of plasma C-reactive protein in biracial (black–white) younger adults: the Bogalusa Heart Study. Metabolism 55:699–705
76. Patel DA, Srinivasan SR, Xu JH, Chen W, Berenson GS (2006) Adiponectin and its correlates of cardiovascular risk in young adults: the Bogalusa Heart Study. Metabolism 55:1551–1557
77. Ridker PM, Wilson PW, Grundy SM (2004) Should C-reactive protein be added to metabolic syndrome and to assessment of global cardiovascular risk? Circulation 109:2818–2825
78. Ford ES, Giles WH, Dietz WH (2002) Prevalence of the metabolic syndrome among US adults: findings from the third National Health and Nutrition Examination Survey. JAMA 287:356–359
79. Ford ES, Giles WH, Mokdad AH (2004) Increasing prevalence of the metabolic syndrome among U.S. adults. Diabetes Care 27:2444–2449
80. Hossain P, Kawar B, El Nahas M (2007) Obesity and diabetes in the developing world—a growing challenge. N Engl J Med 356:213–215
81. Broyles S, Katzmarzyk PT, Srinivasan SR, Chen W, Bouchard C, Freedman DS, Berenson GS (2010)The pediatric obesity epidemic continues unabated in Bogalusa, Louisiana. Pediatrics 125(5):900–905
82. Barker DJ (1998) Mothers, babies and health in later life, 2nd edn. Churchill Livingstone, Edinburgh
83. Frontini MG, Srinivasan SR, Xu J, Berenson GS (2004) Low birth weight and longitudinal trends of cardiovascular risk factor variables from childhood to adolescence: the Bogalusa Heart Study. BMC Pediatr 4:22

Chapter 5
Black–White Divergence Influencing Impaired Fasting Glucose and Type 2 Diabetes Mellitus

Quoc Manh Nguyen, Sathanur R. Srinivasan and Gerald S. Berenson

Abstract Impaired glucose homeostasis is one of the most common causes of death in the U.S. The progressive global epidemic of obesity makes it a major causal factor for pre-diabetes and type 2 diabetes, which may represent the two categories of impaired glucose regulation. Recognition of the importance of black–white contrast in the prevalence of type 2 diabetes mellitus has stirred up interest in its potential role in the development and progression of diabetes. Accumulating evidence suggests that multiple cardiometabolic risk factors such as insulin resistance/hyperinsulinemia, adiposity, genetic predisposition, low birth weight, prepuberty, inflammation markers including C-reactive protein and adiponectin, chronic kidney dysfunction, and environmental and socio-economic status lead to the black–white divergence in type 2 diabetes. Early prevention and intervention for these risk factors, especially obesity and altered lifestyles involving physical activity and dietary consumption, may help to neutralize the racial disparities seen in the emerging epidemic of diabetes beginning in youth.

Keywords Cardio–cerebral–renal metabolism • Environmental factor • Genetic predisposition • Glucose • Impaired fasting glucose • Inflammation marker • Insulin • Insulin resistance • Low birth weight • Metabolic syndrome • Obesity • Pre-diabetes • Puberty • Racial divergence • Socio-economic factor • Type 2 diabetes

5.1 Introduction

Diabetes has become one of the most prevalent chronic disease. It is also associated with a high incidence of mortality in the United States [1]. There are about 19 million people with type 2 diabetes and another 54 million people with impaired fasting glucose or pre-diabetic state in this country [2]. It is widely recognized that diabetes

Q. M. Nguyen (✉)
Department of Epidemiology, Center for Cardiovascular Health, Tulane University School of Public Health and Tropical Medicine, New Orleans, LA, USA
e-mail: qnguyen1@tulane.edu

G. S. Berenson (ed.), *Evolution of Cardio-Metabolic Risk from Birth to Middle Age*,
DOI 10.1007/978-94-007-1451-9_5, © Springer Science+Business Media B.V. 2011

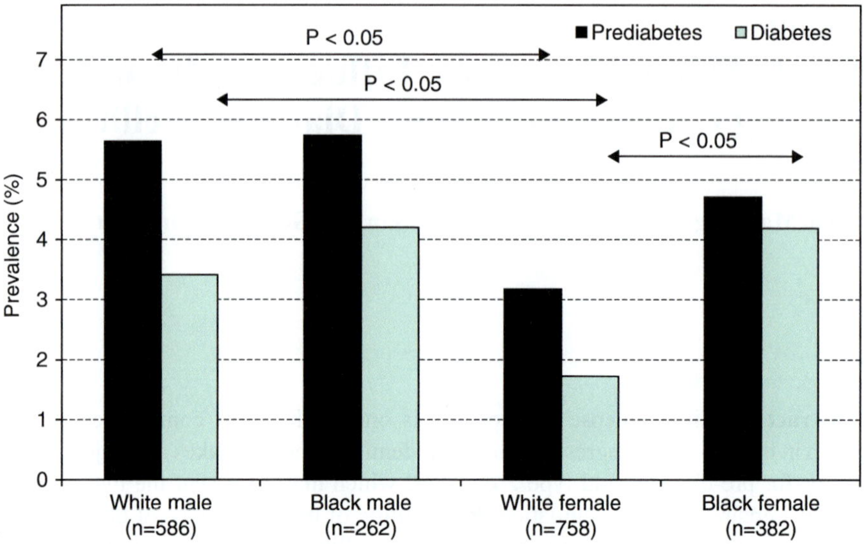

Fig. 5.1 Prevalence of pre-diabetes and diabetes in adulthood, age 19–44, by race and sex. P-values were compared between the race–sex groups. [6]

is a major contributor to adult cardiovascular morbidity and mortality and is a cause of end stage hypertensive-renal disease [3].

Recent epidemiological studies have demonstrated an increasing propensity to developing type 2 diabetes mellitus and its related chronic complications in black–white ethnic groups [2, 4, 5]. Indeed, African–Americans have a greater prevalence of type 2 diabetes, Fig. 5.1, which sets the stage for more long-term complications when compared to Caucasians [2, 4–6]. Further, these observations are made in relatively young individuals. The black–white divergence in prevalence of type 2 diabetes may, in part, be described by the following:

5.2 Insulin Resistance/Hyperinsulinemia

Racial differences in insulin sensitivity are demonstrated beginning in childhood [7]. The Bogalusa Heart Study evaluated plasma glucose and insulin levels during an oral glucose tolerance test in 377 children, aged 5–17 years, from this bi-racial community. After adjusting for weight, age, ponderal index, and pubertal stage, blacks showed higher insulin levels than their white counterparts, which suggests a compensated insulin resistance occurring during early maturation [8]. In addition, blacks, even with higher fasting insulin, have lower C-peptide levels and C-peptide to insulin ratio than their white counterparts. Such observations indicate a lower insulin secretion by pancreatic beta cells and a reduced hepatic insulin clearance in blacks at the adolescent age (Fig. 5.2) [9]. Osei et al. have also reported that

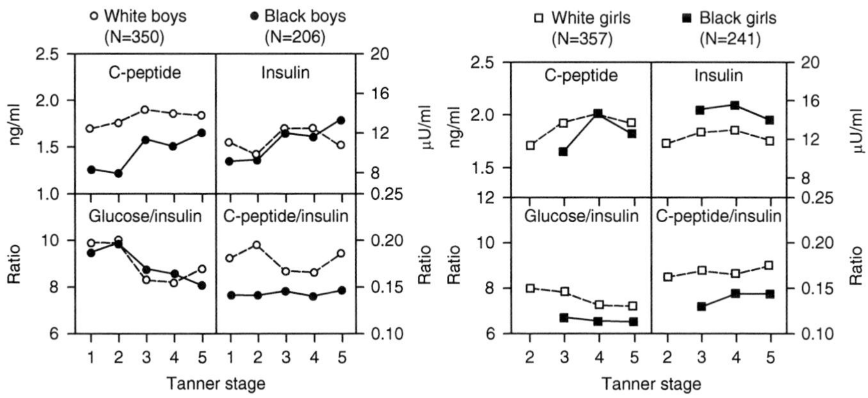

Fig. 5.2 Mean levels of metabolic syndrome parameters based on sexual maturation and race in boys and girls. C-peptide secretion is shown to be less in blacks. [9]

there are ethnic differences (black–white) in the regulation of glucose homeostasis as well as insulin secretion, action, and metabolism [4]. Decreased hepatic insulin extraction and clearance in blacks and the resultant peripheral hyperinsulinemia implicates peripheral insulin insensitivity (due to downregulation of the tissue insulin receptors) and the contribution to development of obesity and future type 2 diabetes in blacks [4]. Further, non-obese prepubertal black children can have decreased insulin sensitivity and lower adiponectin levels compared to white children [10, 11]. These findings suggest that black children may have a genetic predisposition to insulin resistance. In the presence of environmental modulators, their risk of type 2 diabetes increases and leads to disease expression during physiologic (puberty) or pathologic (obesity) states of insulin resistance [7].

5.3 Obesity

Recent United States national data indicate the growing obesity epidemic and racial disparities continue to increase in children being overweight [12]. Previous findings show obesity is more common in blacks than in whites, and is still increasing, especially in black adolescents (Fig. 5.3) [13, 14]. In African–American children, as body mass index (BMI) increases, insulin-stimulated glucose metabolism decreases and insulin levels increase [7, 15]. In addition, the inverse relationship between insulin sensitivity and abdominal fat is stronger for visceral than for subcutaneous fat [7]. Despite 30% lower visceral adiposity in black adolescents, insulin sensitivity is not better than that of their white peers [16, 17]. Moreover, glycated hemoglobin (HbA_{1c}), an indicator of long-term glucose homeostasis, is positively correlated with waist circumference, a measure of central body fatness related to metabolic syndrome, independent of race, insulin resistance and other cardiometabolic risk factors in nondiabetic younger populations [18].

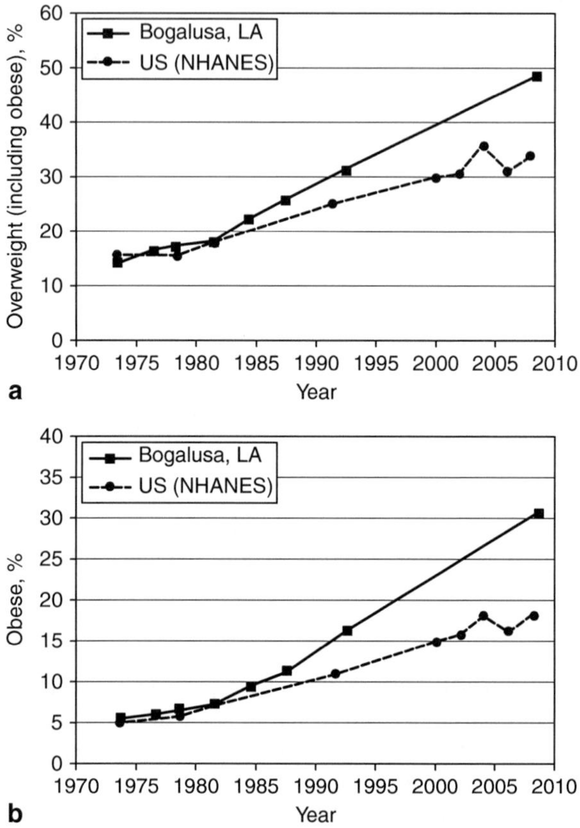

Fig. 5.3 Bogalusa, LA, compared with the United States (NHANES): changes in the proportion of children and adolescents 5–17 years of age classified as being overweight (BMI≥85th percentile; includes obese) (**a**) or obese (BMI≥95th percentile) (**b**). Recent data collected in 2009. [14]

In different populations, studies show obesity in young children and adolescents is a strong predictor of subsequent diabetes [6, 19]. Because obesity is pathologically linked to insulin resistance/hyperinsulinemia, it plays a crucial role as an initiating factor in the development of dysglycemia [20]. This is consistent with observations showing temporal associations between the degree of baseline adiposity and later incidence of hyperinsulinemia [20] or metabolic syndrome [21], independent of baseline insulin levels (Fig. 5.4). Metabolically, initially obesity precedes hyperinsulinemia. Insulin later drives obesity, probably through multiple hormonal interactions. It is well-known that insulin increases glycogen storage [22], drives appetite, interacts with leptin [22, 23], and has strong growth characteristic [24].

5.4 Genetic Predisposition

There is a strong hereditary (multigenic) component to diabetes, with the role of genetic determinants demonstrated when black–white divergences in the prevalence of type 2 diabetes occur [7]. Earlier studies using genetic admixture analy-

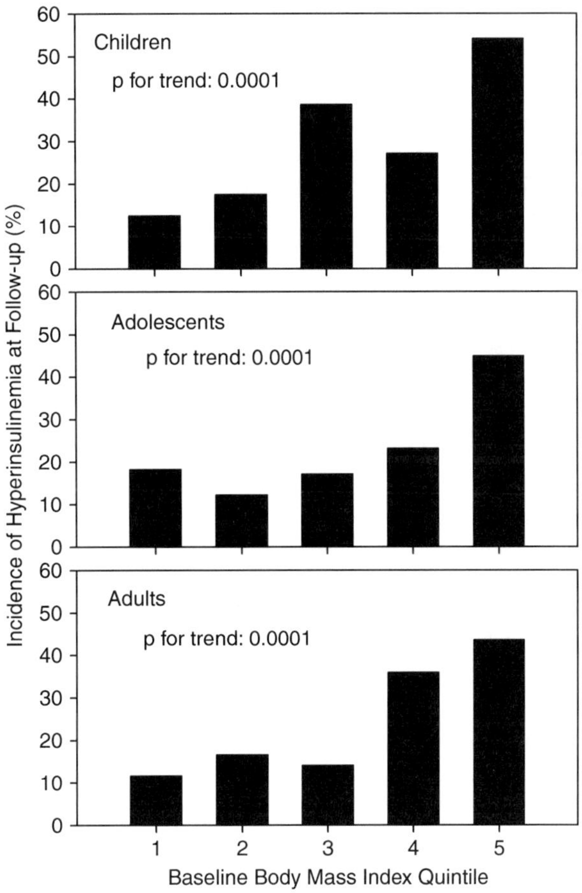

Fig. 5.4 Incidence of hyperinsulinemia (insulin > 75th percentile, specific for age, race, gender, and survey year) at follow-up study (after 3 years) in children, adolescents, and adults by body mass index quintiles (specific for age, race, gender, and survey year) at baseline. [20]

sis suggest a genetic and environmental basis to these divergences [16, 25]. In countries characterized with thin individuals in the population, individuals still have a diabetic propensity indicative of the "thrifty gene concept" [26–28]. The Japanese-Honolulu-San Francisco (NiHon San) studies, with individuals having the same genetic background, show Japanese immigrating to a westernized environment develop more diabetes [29, 30]. Although the genetic relation to diabetes is obviously very complex and related to multiple genes, these studies clearly show the environmental-genetic interrelationships. Our report on β-adrenergic genes is consistent with racial differences that occur [31]. This study shows the Gly16 allele was more frequent and the Arg16 allele was less frequent in whites relative to blacks. Further, parental history being a strong characteristic for future diabetes in offspring implicates both a genetic background as well as an environmental input [32].

5.5 Low Birth Weight

Previous reports show birth weights are lower among blacks and those living in the southeast region of the United States [33, 34]. It is now well-known that the mismatch of a poor intrauterine environment and a nutritionally rich environment later in life increases the risk of several noncommunicable diseases, such as coronary heart disease and diabetes [35, 36]. As a result, the predisposition to type 2 diabetes may be enhanced and even programmed in utero [27, 33, 34]. Low birth weight, representing the effects of intrauterine undernutrition, is a risk factor for insulin resistance. Also in utero exposure to gestational diabetes has been found to decrease insulin secretory capacity [37] and predicts diabetes later in life [10, 38, 39].

5.6 Puberty

Puberty appears to play a major role in the development of type 2 diabetes [7]. Prepubertal black children have lower resting energy expenditure and higher fasting and first phase insulin concentrations noted with glucose clamp studies than did white youngsters [40]. Among pubertal individuals, blacks also have lower insulin sensitivity [41]. The lower resting energy expenditure in black children is suggested to promote a greater susceptibility to obesity [40]. Environmentally and by lifestyles, this may be particularly true in black females who begin to show extreme obesity with an onset of around nine years of age [42]. Our publication demonstrated that girls at the 85th percentile or above of body weight showed black girls increasing their BMI faster than white girls and a greater adverse increase of CV risk factors [43].

5.7 Inflammation Markers

The levels of the anti-diabetogenic hormone adiponectin are lower in blacks compared with white healthy children with similar body composition [10, 11, 16, 44, 45]. In a similar population environment, blacks also had lower adiponectin levels than whites in the Bogalusa Heart Study [45]. High-sensitivity C-reactive protein (hCRP) levels are also found to be greater in girls and greater in black females [46, 47]. Although hCRP is not a cause of obesity, it reflects the adverse change occurring with beginning lipotoxicity and related systemic inflammation [46, 47].

5.8 Glucose–Insulin Homeostasis Metabolism Potentially Influences Cardio–Cerebral–Renal Metabolism

Diabetes and hypertensive vascular disease cause 60–75% of end-stage renal disease (ESRD) worldwide [48, 49]. Chronic kidney disease (CKD) is a strong independent risk factor for CV disease, and, in turn, this becomes the primary complication and cause of death among CKD patients [49, 50]. Further, CV mortality due to CKD earliest stages is around 15 times greater in dialysis patients than in the general population [51, 52]. The early evidence of abnormal risk factors in pre-diabetes and diabetes in young individuals reflects the "silent" effects on the cardiovascular renal system [6].

Studies of microalbuminuria in blacks are indication of blacks propensity to develop diabetes and renal damage. Positive microalbuminuria is not only a marker of renal structural changes, but any increase of albumin in urine implies systemic vascular damage [49, 53]. The kidney is the only vascularized organ that communicates with the exterior through fluid filtered through its microvasculature [49, 53]. Because of its significance, CKD becomes a tracer for other diseases, like hypertension, that produce early damage to the cardiovascular system, especially in black males [49, 53, 54].

Given the global epidemic increase in the incidence of the leading causes of CKD such as hypertension, obesity and diabetes mellitus [55, 56], primary prevention and early detection of CKD are obviously needed. Effective intervention strategies are available to slow the progression of CKD and reduce cardiovascular risk, increasing the importance of its detection, especially in blacks [51, 57].

5.9 Childhood Progression to Type 2 Diabetes (Glucose Levels, Parental History of Diabetes and Obesity)

The deterioration in glucose levels to pre-diabetes or diabetes that, after a relatively stable period, occurs as a rapid, incremental increase accompanied by a decline in insulin sensitivity [58]. However, in the Bogalusa Heart Study, both pre-diabetes and diabetes showed progressively increasing glucose and adiposity levels beginning in early life, before the onset of impaired glucose regulation status, suggesting that even small changes in fasting glucose levels may be a marker of altered carbohydrate–insulin imbalance [6].

In addition, parental history of diabetes observed in childhood is the most consistent predictor of adverse changes leading to impaired glucose regulation status, regardless of age and other risk factors (Fig. 5.5) [32]. A number of studies have shown parental diabetes as an independent risk factor for type 2 diabetes [19, 59, 60]. Further, the prevalence of excess HbA_{1c} is significantly associated with a positive parental history of diabetes and type 2 diabetes mellitus [18].

5.10 Environmental and Socio-Economic Factors

Race disparities in diabetes also stem from divergences in the health risk environment that blacks and whites live, like lower socio-economic lifestyles, access to food, and nature of food consumption, etc… When blacks and whites live in similar risk environment, their health outcomes become more similar [61].

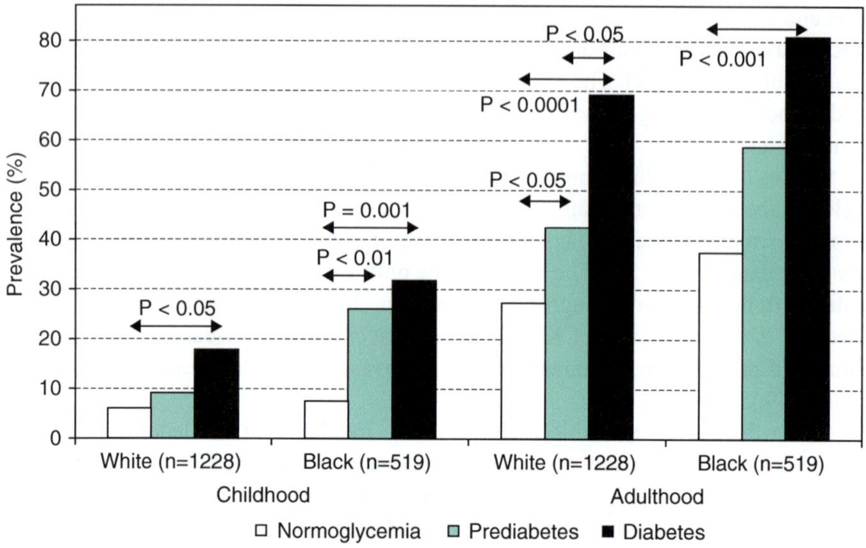

Fig. 5.5 Prevalence of parental history of diabetes in childhood and adulthood by race, prediabetic and diabetic status in adulthood. Significant differences in the prevalence of parental history of diabetes by race and diabetes status were tested by the Pearson's χ^2-test or Fisher's exact test, as applicable. [32]

5.11 Conclusion

Multiple cardiometabolic risk factors and lifestyles are evident in black–white divergences related to type 2 diabetes. Early prevention and intervention for risk factors, particularly obesity and altered lifestyles involving physical activity and dietary consumption, will help to counteract the racial disparities seen in the emerging epidemic of diabetes beginning in youth in our population.

References

1. Hoyert DL, Heron MP, Murphy SL, Kung H (2006) Deaths. Final data for 2003. National vital statistics reports, vol 54, No 13. National Center for Health Statistics, Hyattsville
2. Cowie CC, Rust KF, Byrd-Holt DD, Eberhardt MS, Flegal KM, Engelgau MM, Saydah SH, Williams DE, Geiss LS, Gregg EW (2006) Prevalence of diabetes and impaired fasting glucose in adults in the U.S. population: National Health And Nutrition Examination Survey 1999–2002. Diabetes Care 29:1263–1268
3. Fox CS, Sullivan L, D'Agostino RB Sr, Wilson PW. (2004) Framingham Heart Study. The significant effect of diabetes duration on coronary heart disease mortality: the Framingham Heart Study. Diabetes Care 27:704–708
4. Osei K, Schuster DP (1994) Ethnic differences in secretion, sensitivity, and hepatic extraction of insulin in black and white Americans. Diabet Med 11:755–62

5. Zimmet P (1982) Type 2 (non-insulin-dependent) diabetes—an epidemiological overview. Diabetologia 22:399–411
6. Nguyen QM, Srinivasan SR, Xu JH, Chen W, Berenson GS (2008) Changes in risk variables of metabolic syndrome since childhood in pre-diabetic and type 2 diabetic subjects: the Bogalusa Heart Study. Diabetes Care 31:2044–2049
7. Richard Kahn (2000) American Diabetes Association. Type 2 diabetes in children and adolescents. Diabetes Care 23:381–389
8. Svec F, Nastasi K, Hilton C, Bao W, Srinivasan SR, Berenson GS. (1992) Black–white contrasts in insulin levels during pubertal development. The Bogalusa Heart Study. Diabetes 41:313–317
9. Jiang X, Srinivasan SR, Radhakrishnamurthy B, Dalferes ER, Berenson GS (1996) Racial (black–white) differences in insulin secretion and clearance in adolescents: the Bogalusa Heart Study. Pediatrics 97:357–360
10. Botero D, Wolfsdorf JI (2005) Diabetes mellitus in children and adolescents. Arch Med Res 36:281–290
11. Arslanian SA, Saad R, Lewy V, Danadian K, Janosky J (2002) Hyperinsulinemia in African–American children: decreased insulin clearance and increased insulin secretion and its relationship to insulin sensitivity. Diabetes 51:3014–3019
12. Strauss RS, Pollack HA (2001) Epidemic increase in childhood overweight, 1986–1998. JAMA 286:2845–2848
13. Okosun IS (2001) Racial differences in rates of type 2 diabetes in American women: how much is due to differences in overall adiposity? Ethn Health 6:27–34
14. Broyles S, Katzmarzyk PT, Srinivasan SR, Chen W, Bouchard C, Freedman DS, Berenson GS (2010) The pediatric obesity epidemic continues unabated in Bogalusa, Louisiana. Pediatrics 125:900–905
15. Schuster DP, Kien CL, Osei K (1998) Differential impact of obesity on glucose metabolism in black and white American adolescents. Am J Med Sci 316:361–367
16. Gungor N, Hannon T, Libman I, Bacha F, Arslanian S (2005) Type 2 diabetes mellitus in youth: the complete picture to date. Pediatr Clin North Am 52:1579–1609
17. Bacha F, Saad R, Gungor N, Janosky J, Arslanian SA (2003) Obesity, regional fat distribution, and syndrome X in obese black versus white adolescents: race differential in diabetogenic and atherogenic risk factors. J Clin Endocrinol Metab 88:2534–2540
18. Nguyen QM, Srinivasan SR, Xu JH, Chen W, Berenson GS (2008) Distribution and cardiovascular risk correlates of hemoglobin A(1c) in nondiabetic younger adults: the Bogalusa Heart Study. Metabolism 57:1487–1492
19. Franks PW, Hanson RL, Knowler WC, Moffett C, Enos G, Infante AM, Krakoff J, Looker HC (2007) Childhood predictors of young-onset type 2 diabetes. Diabetes 56:2964–2972
20. Srinivasan SR, Myers L, Berenson GS (1999) Temporal association between obesity and hyperinsulinemia in children, adolescents, and young adults: the Bogalusa Heart Study. Metabolism 48:928–934
21. Srinivasan SR, Myers L, Berenson GS (2002) Predictability of childhood adiposity and insulin for developing insulin resistance syndrome (syndrome X) in young adulthood: the Bogalusa Heart Study. Diabetes 51:204–209
22. Thorens B (2008) Glucose sensing and the pathogenesis of obesity and type 2 diabetes. Int J Obes (Lond) 32(Suppl 6):62–71
23. Sonnett TE, Levien TL, Gates BJ, Robinson JD, Campbell RK (2010) Diabetes mellitus, inflammation, obesity: proposed treatment pathways for current and future therapies. Ann Pharmacother 44:701–711
24. Møller N, Jørgensen JO (2009) Effects of growth hormone on glucose, lipid, and protein metabolism in human subjects. Endocr Rev 30:152–177
25. Gower BA, Fernández JR, Beasley TM, Shriver MD, Goran MI (2003) Using genetic admixture to explain racial differences in insulin-related phenotypes. Diabetes 52:1047–1051
26. Carulli L, Rondinella S, Lombardini S, Canedi I, Loria P, Carulli N (2005) Review article: diabetes, genetics and ethnicity. Aliment Pharmacol Ther 22(Suppl 2):16–19

27. Stocker CJ, Arch JR, Cawthorne MA (2005) Fetal origins of insulin resistance and obesity. Proc Nutr Soc 64:143–151
28. Paradies YC, Montoya MJ, Fullerton SM (2007) Racialized genetics and the study of complex diseases: the thrifty genotype revisited. Perspect Biol Med 50:203–227
29. Huang B, Rodriguez BL, Burchfiel CM, Chyou PH, Curb JD, Yano K (1996) Acculturation and prevalence of diabetes among Japanese–American men in Hawaii. Am J Epidemiol 144:674–681
30. Rodriguez BL, Curb JD, Burchfiel CM, Huang B, Sharp DS, Lu GY, Fujimoto W, Yano K (1996) Impaired glucose tolerance, diabetes, and cardiovascular disease risk factor profiles in the elderly. The Honolulu Heart Program. Diabetes Care 19:587–590
31. Ellsworth DL, Coady SA, Chen W, Srinivasan SR, Elkasabany A, Gustat J, Boerwinkle E, Berenson GS (2002) Influence of the beta2-adrenergic receptor Arg16Gly polymorphism on longitudinal changes in obesity from childhood through young adulthood in a biracial cohort: the Bogalusa Heart Study. Int J Obes Relat Metab Disord 26:928–937
32. Nguyen QM, Srinivasan SR, Xu JH, Chen W, Berenson GS (2009) Influence of childhood parental history of type 2 diabetes on the pre-diabetic and diabetic status in adulthood: the Bogalusa Heart Study. Eur J Epidemiol 24:537–539
33. Lackland DT, Egan BM, Syddall HE, Barker DJ (2002) Associations between birth weight and antihypertensive medication in black and white medicaid recipients. Hypertension 39:179–183
34. Ventura SJ, Martin JA, Curtin SC, Mathews TJ, Park MM (2000) National vital statistics report. U.S. Department of Health and Human Services, Hyattsville, pp 1–100
35. Eriksson JG, Forsén TJ (2002) Childhood growth and coronary heart disease in later life. Ann Med 34:157–161
36. Eriksson JG, Forsén T, Tuomilehto J, Jaddoe VW, Osmond C, Barker DJ (2002) Effects of size at birth and childhood growth on the insulin resistance syndrome in elderly individuals. Diabetologia 45:342–348
37. Gautier JF, Wilson C, Weyer C, Mott D, Knowler WC, Cavaghan M, Polonsky KS, Bogardus C, Pratley RE (2001) Low acute insulin secretory responses in adult offspring of people with early onset type 2 diabetes. Diabetes 50:1828–1833
38. Forsén T, Eriksson J, Tuomilehto J, Reunanen A, Osmond C, Barker D (2000) The fetal and childhood growth of persons who develop type 2 diabetes. Ann Intern Med 133:176–182
39. Whincup PH, Kaye SJ, Owen CG, Huxley R, Cook DG, Anazawa S, Barrett-Connor E, Bhargava SK, Birgisdottir BE, Carlsson S, de Rooij SR, Dyck RF, Eriksson JG, Falkner B, Fall C, Forsén T, Grill V, Gudnason V, Hulman S, Hyppönen E, Jeffreys M, Lawlor DA, Leon DA, Minami J, Mishra G, Osmond C, Power C, Rich-Edwards JW, Roseboom TJ, Sachdev HS, Syddall H, Thorsdottir I, Vanhala M, Wadsworth M, Yarbrough DE (2008) Birth weight and risk of type 2 diabetes: a systematic review. JAMA 300:2886–2897
40. Morrison JA, Alfaro MP, Khoury P, Thornton BB, Daniels SR (1996) Determinants of resting energy expenditure in young black girls and young white girls. J Pediatr 129:637–642
41. Arslanian S, Suprasongsin C (1996) Differences in the in vivo insulin secretion and sensitivity of healthy black versus white adolescents. J Pediatr 129:440–443
42. Webber LS, Cresanta JL, Croft JB, Srinivasan SR, Berenson GS (1986) Transitions of cardiovascular risk from adolescence to young adulthood—the Bogalusa Heart Study: II. Alterations in anthropometric blood pressure and serum lipoprotein variables. J Chronic Dis 39:91–103
43. Freedman DS, Dietz WH, Srinivasan SR, Berenson GS (1999) The relation of overweight to cardiovascular risk factors among children and adolescents: the Bogalusa Heart Study. Pediatrics 103:1175–1182
44. Bacha F, Saad R, Gungor N, Arslanian SA (2005) Does adiponectin explain the lower insulin sensitivity and hyperinsulinemia of African–American children? Pediatr Diabetes 6:100–102
45. Patel DA, Srinivasan SR, Xu JH, Chen W, Berenson GS (2006) Adiponectin and its correlates of cardiovascular risk in young adults: the Bogalusa Heart Study. Metabolism 55:1551–1557

46. Toprak D, Toprak A, Chen W, Xu JH, Srinivasan S, Berenson GS (2011) Adiposity in childhood is related to C-reactive protein and adiponectin in young adulthood: from the Bogalusa Heart Study. Obesity (Silver Spring) 19:185–190
47. Patel DA, Srinivasan SR, Xu JH, Li S, Chen W, Berenson GS. (2006) Distribution and metabolic syndrome correlates of plasma C-reactive protein in biracial (black–white) younger adults: the Bogalusa Heart Study. Metabolism 55:699–705
48. U.S. Renal Data Systems USRDS 2007 (2010) Annual data report: Atlas of end stage renal disease in the United States. National Institutes of Health, NI-DDK, Bethesda, MD, 2007. Incidence & prevalence. Am J Kidney Dis 55(Suppl 1):S231–S240
49. Herrera R, Almaguer M, Chipi J, Martínez O, Bacallao J, Rodríguez N, Abreu Mde L, Fariña O, Roche Mdel C (2010) Albuminuria as a marker of kidney and cardio–cerebral vascular damage. Isle of Youth Study (ISYS), Cuba. MEDICC Rev 12:20–26
50. Chauveau P, Rigalleau V, Aparicio M (2008) Insulin resistance and chronic kidney disease. Nephrol Ther 4:568–574
51. Bello AK, Nwankwo E, El Nahas AM(2005) Prevention of chronic kidney disease: a global challenge. Kidney Int 98(Suppl):S11–S17
52. Sarnak MJ, Levey AS (2000) Cardiovascular disease and chronic renal disease: a new paradigm. Am J Kidney Dis 35(4 Suppl 1):S117–S131
53. de Zeeuw D (2004) Albuminuria, not only a cardiovascular/renal risk marker, but also a target for treatment? Kidney Int 92(Suppl):S2–S6
54. Muntner P, Arshad A, Morse SA, Patel DA, Manapatra PD, Reisin E, Aguilar EA, Chen W, Srinivasan S, Berenson GS (2009) End-stage renal disease in young black males in a black–white population: longitudinal analysis of the Bogalusa Heart Study. BMC Nephrol 10:40
55. Naser KA, Gruber A, Thomson GA (2006) The emerging pandemic of obesity and diabetes: are we doing enough to prevent a disaster? Int J Clin Pract 60:1093–1097
56. Cowie CC, Rust KF, Ford ES, Eberhardt MS, Byrd-Holt DD, Li C, Williams DE, Gregg EW, Bainbridge KE, Saydah SH, Geiss LS (2009) Full accounting of diabetes and pre-diabetes in the U.S. Population in 1988–1994 and 2005–2006. Diabetes Care 32:287–294
57. Stenvinkel P (2010) Chronic kidney disease: a public health priority and harbinger of premature cardiovascular disease. J Intern Med 268:456–467
58. Laspa E, Christen A, Efstathiadou Z, Johnston DG, Godsland IF (2007) Long-term changes and variability in diabetes risk factors prior to the development of impaired glucose homeostasis. Diabet Med 24:1269–1278
59. Klein BE, Klein R, Moss SE, Cruickshanks KJ (1996) Parental history of diabetes in a population-based study. Diabetes Care 19:827–830
60. Wilson PW, Meigs JB, Sullivan L, Fox CS, Nathan DM, D'Agostino RB Sr (2007) Prediction of incident diabetes mellitus in middle aged adults: the Framingham offspring study. Arch Intern Med 167:1068–1074
61. LaVeist TA, Thorpe RJ Jr, Galarraga JE, Bower KM, Gary-Webb TL (2009) Environmental and socio-economic factors as contributors to racial disparities in diabetes prevalence. J Gen Intern Med 24:1144–1148

Chapter 6
Birth Weight, Stimulus Response and Hemodynamic Variability Implicate Racial (Black–White) Contrasts of Autonomic Control of Heart Rate and Blood Pressure and Related Cardiovascular Disease

Gerald S. Berenson, Pronabesh DasMahapatra, Camilo Fernandez Alonso, Wei Chen, Jihua Xu, Thomas Giles and Sathanur R. Srinivasan

Abstract The mechanisms controlling heart rate and blood pressure (BP) are quite complex. To some extent these hemodynamic variables are interrelated but observations from the Bogalusa Heart Study suggest different mechanisms are involved in their control and are influenced by ethnicity and gender. Heart rate seems to have a greater central autonomic control, being slower in black children than in white children, while risk factors, heart rate and BP are less associated with body fatness in blacks. Heart rate variability indicates a greater vagal control in blacks. The response of heart rate to stressors is greater in white children while the response of BP to various stimuli is greater in black children. Low birth weight is also associated with long-term BP variability. Also, the long term variability of BP is greater in blacks and is associated with development of left ventricular hypertrophy, evidence of variability having a greater impact on the cardiovascular system. Based on clinical observations BP obviously involves many mechanisms, i.e. arterial wall structure, endothelial function, nitric oxide production, the renin-angiotensin system, electrolytes and other. Although epidemiologic observations do not establish such mechanisms, they have implications of their existence. The racial contrasts help elucidate such mechanisms and help guide prevention strategies.

Keywords Heart rate • Blood pressure • Control mechanism for heart rate • Control mechanism for blood pressure • Autonomic control of hemodynamic characteristics • Black–white gender contrast • Prevention

---

G. S. Berenson (✉)
Department of Medicine, Pediatrics, Biochemistry, Epidemiology, Center for Cardiovascular Health, Tulane University School of Medicine and School of Public Health and Tropical Medicine, New Orleans, LA, USA
e-mail: berenson@tulane.edu

G. S. Berenson (ed.), *Evolution of Cardio-Metabolic Risk from Birth to Middle Age*,
DOI 10.1007/978-94-007-1451-9_6, © Springer Science+Business Media B.V. 2011

6.1 Introduction

Heart rate variability and etiologies of essential hypertension interact. These hemodynamic variables are multifactorial in nature involving an interplay of genes and a host of environmental factors and lifestyles. Marked race (black–white) and sex differences in both heart rate and blood pressure (BP) have been reported in childhood. Yet, despite decades of research, specific genetic and pathophysiologic mechanisms mediating the hemodynamic variables are not fully understood [1].

BP is a highly variable physiologic trait that increases with age and body growth from childhood, while heart rate tends to decrease into adulthood. The association between birth weight and BP also increases related to aging. There are differences in BP levels among individuals at a given time (among-individual variability) and both short-term and long-term fluctuations occur within the same individual at different time points (within-individual variability) [2, 3]. Studies have shown that wide variability of BP over 24 h or over a long-term period (years) are associated with severity of end organ damage and an increase of subsequent cardiovascular (CV) and cerebrovascular events, even after adjusting for mean BP levels [4, 5]. Limited studies concern variability over long periods and a response to stimuli (stress) as they influence heart disease.

Recently, long-term BP variability, as well as 24 h monitoring, has received increasing attention because of its implications for treatment and prevention of target organ damage [4–11]. Long-term BP variability, especially beginning in childhood, can be a more powerful determinant than average levels in predicting cardiac disease. We have found this to occur particularly in blacks compared to white individuals. Comparative biracial (black–white) studies in this regard provided an opportunity to study variations in and among individuals. These observations also provide clues to mechanisms controlling heart rate and BP with important implications for treatment and prevention.

6.2 Studies in Early Life

The relation between low birth weight, an indicator of intrauterine growth retardation, and elevated BP levels has become well known [12–17]. The magnitude of birth weight—BP association was found to increase with age from childhood to adulthood [18–20]. Further study of the age amplification was supported by observations in Bogalusa. Individuals (N=6251) with available birth weight were examined 1–15 times for BP from childhood to adulthood. The magnitude of the birth weight—systolic BP relationship was significantly amplified with increasing age, even adjusting for body mass index (BMI) and race (Fig. 6.1). The magnitude of the inverse association was considerably related to the strength of birth weight- current weight and current weight—BP correlations, while adjustment for current BMI and height even strengthened the negative association [21]. Low birth weight was also correlated with greater variability of BP in adulthood [22]

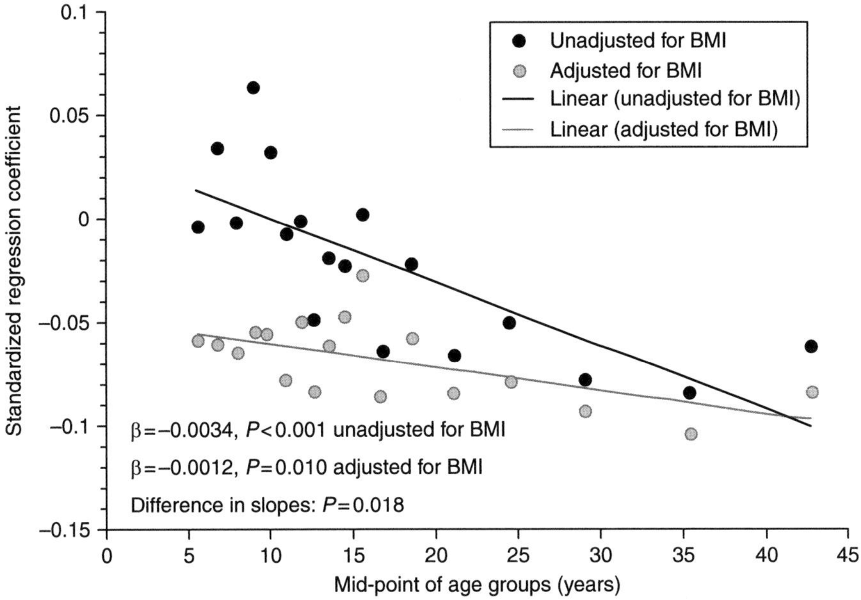

Fig. 6.1 Age-related trend of the relationship between birth weight and systolic BP with or without adjustment for current BMI by age groups. [21]

Birth weight has been found to be associated with subclinical cardiovascular system changes, measured by brachial-ankle pulse wave velocity (baPWV), a marker of arterial stiffness [23]. Besides baPWV, birth weight was also found to correlate inversely with pulse pressure and systolic and diastolic BP. Such changes suggest, that although baPWV and pulse pressure are indicators of arterial stiffness, baPWV may be more closely related to changes of the arterial tree, like elastic properties of the aorta and larger arterial branches, while pulse pressure is also affected by characteristics of the reflected pressure wave from peripheral and more resistant vessels. These studies thus provide some evidence relating birth weight to later hemodynamic measures and even structural changes in the cardiovascular system.

Heart rate and blood pressure were explored in relation to arterial wall stiffness and thickness, and cardiac size. Adjusting for traditional cardiovascular risk variables, heart rate was independently associated with afPWV but not with carotid intima-media thickness (IMT) indicating heart rate plays a different role in developing arterial stiffness and subclinical atherosclerosis in young individuals [24]. Different effects of risk factors were found for PWV from carotid IMT when related to non-HDL cholesterol compared to systolic BP (Fig. 6.2). PWV was more related to systolic BP while IMT more to non-HDL cholesterol. Other studies using heart rate variability of low to high frequency power and time domain PNN50 along with various reactivity tests indicated a greater parasympathetic activity in black children and more sympathetic activity in white children and an increased sympathetic activity in black children in those with higher BP levels [25]. Thus, heart rate compared to BP may be relatively more dependent on autonomic control.

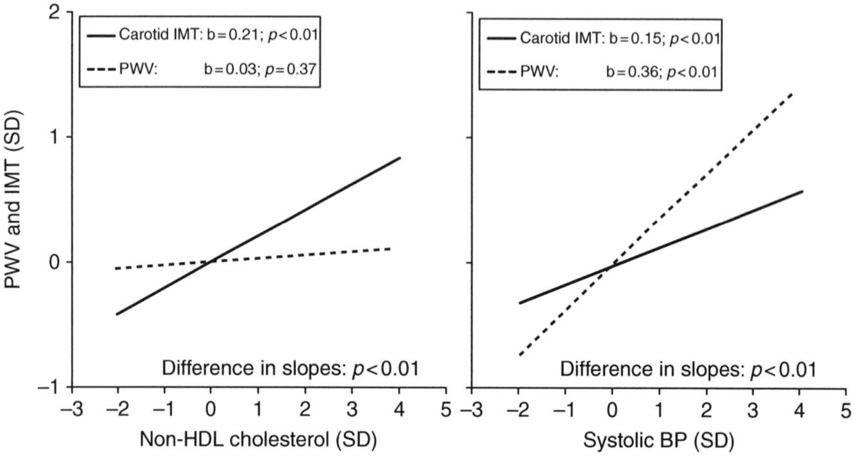

Fig. 6.2 Relationship of non-HDL cholesterol and systolic BP to afPWV and carotid IMT. [24]

With respect to cardiac structure and function, left ventricular wall thickness related to systolic BP, even when adjusted for body surface area, ponderosity (wt/ht^3), race and sex [26]. These observations, along with others, indicated increasing left ventricular thickness in the top quintile of systolic BP levels, Fig. 6.3, consistent with BP related to cardiac size even beginning in childhood, but importantly found at BP levels considered within normal limits. Also, measures of cardiac output were greater in white boys than black boys. The reverse occurs with blacks showing a greater peripheral resistance consistent with the known higher blood pressure levels in black children as a prelude to hypertension in adults [27].

In a Special Blood Pressure Study of children stratified by low, mid and high (1–5 strata) levels of BP, interesting racial contrasts in the complexity of variables relating to BP and heart rate were noted, as shown in Table 6.1 [28–30]. In all BP strata, blacks had higher BP, while white children had greater body fat, faster heart rates, greater plasma renin activity and dopamine β hydroxylase (DBH) levels, higher fasting plasma glucose and greater 24 h urine K^+. In the high BP (strata 5)

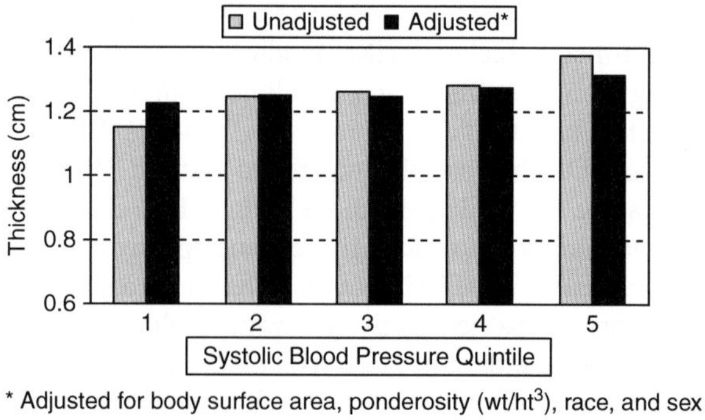

Fig. 6.3 Left ventricular wall thickness in systole across systolic BP quintile groups. [26]

Table 6.1 Contrasts regarding correlates of blood pressure levels in black and white children. [29]

White > Blacks	Blacks > Whites
All blood pressure strata	
Percentage body fat	Blood pressure (BP)
Plasma renin activity	Lower urine K^+ excretion
Dopamine β hydroxylase (DβH)	
Fasting plasma glucose	
High blood pressure strata	
Resting heart rate	Peripheral resistance
Cardiac output	Correlation of 24 h urine Na^+ with BP (males only)
Renin activity and DβH	Inverse correlation of plasma renin with BP
1 h postprandial plasma glucose	

peripheral resistance was greater in black children and a negative correlation with plasma renin and higher correlation with 24 h urine Na^+. In the high BP strata white children had higher resting heart rate, higher plasma renin and DBH and 1 h post prandial plasma glucose. Overall, these observations suggested more sympathetic control of heart rate and BP levels in white children and a greater role of electrolytes and renin-angiotensin-aldosterone system control of BP in their black counterparts. The slower resting heart rate and the heart rate variability studies showed more parasympathetic activity in blacks vs whites.

Further studies of subjects stratified by BP levels were conducted at rest with respect to BP response to orthostatic changes, response to hand grip isometric changes and to the cold pressure test (Fig. 6.4). The trend of stimuli from reactivity tests of hand grip and cold pressor on blood pressure was greater in black children suggesting a shift in sympathetic/parasympathetic control during stress and stimulus response. The BP responses were greater in blacks, especially in males. Of interest, the opposite occurred in white children with a greater increase in heart rate to stress responses (Fig. 6.4). These contrasting stimulus responses are consistent with different mechanisms operative in BP control from heart rate regulation, as discussed later. Overall, these observations show the complexity of heart rate and BP control and implicate multiple mechanisms operating in childhood.

6.3 Blood Pressure and Heart Rate Studies in Young Adults

Studies published by Coronary Artery Risk Development in Young Adults [31] indicated black adults have higher heart rates than their white counterparts. Other studies suggested greater sympathetic activity by peroneal nerve studies in blacks [32, 33]. These observations were contrary to our earlier findings of slower heart rates in black children and heart rate variability reflecting more parasympathetic activity in young black adults [25, 30, 34]. As expected, blood pressure changes from child-

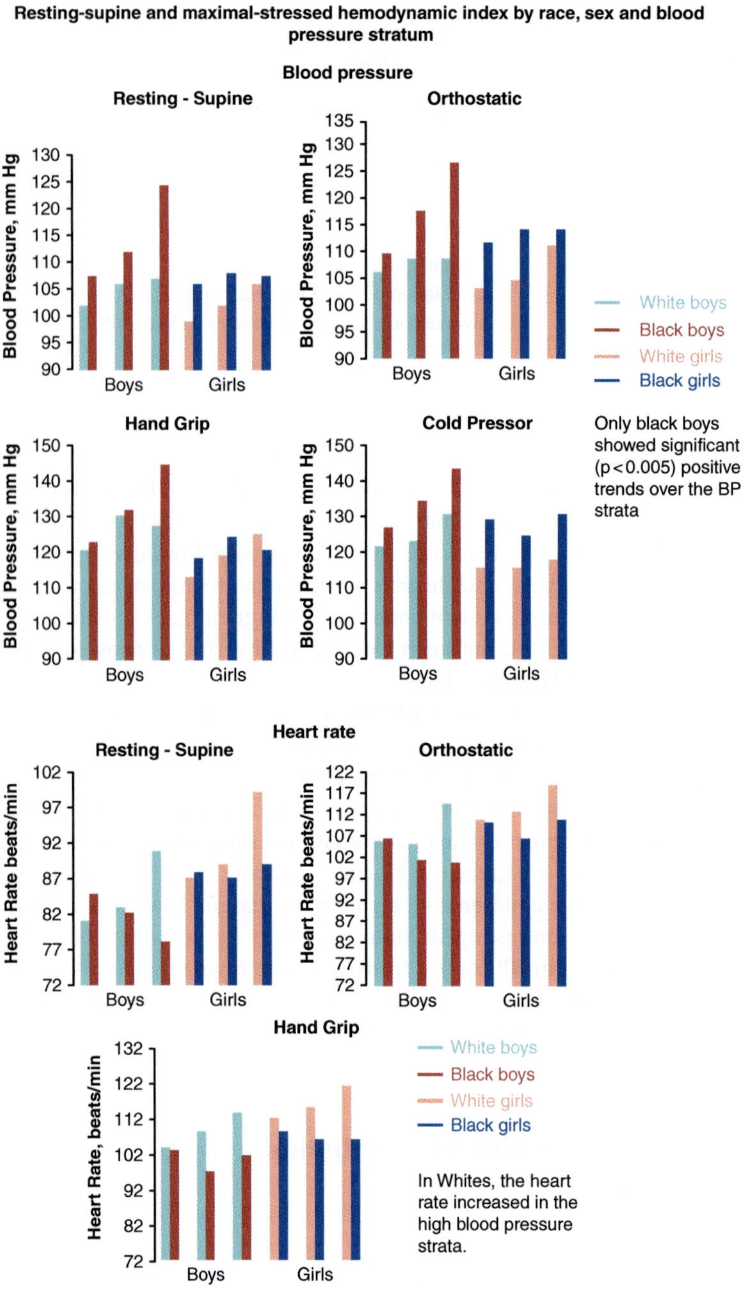

Fig. 6.4 Stimulus response to various conditions shown by resting-supine and maximal-stressed systolic BP as measured by the Whittaker Sphygmostat, by race, sex, and BP stratum. Black boys showed most significant ($p<0.005$) positive trends over the BP strata for a variety of stresses. In contrast, white children showed greater response for heart rate. [30]

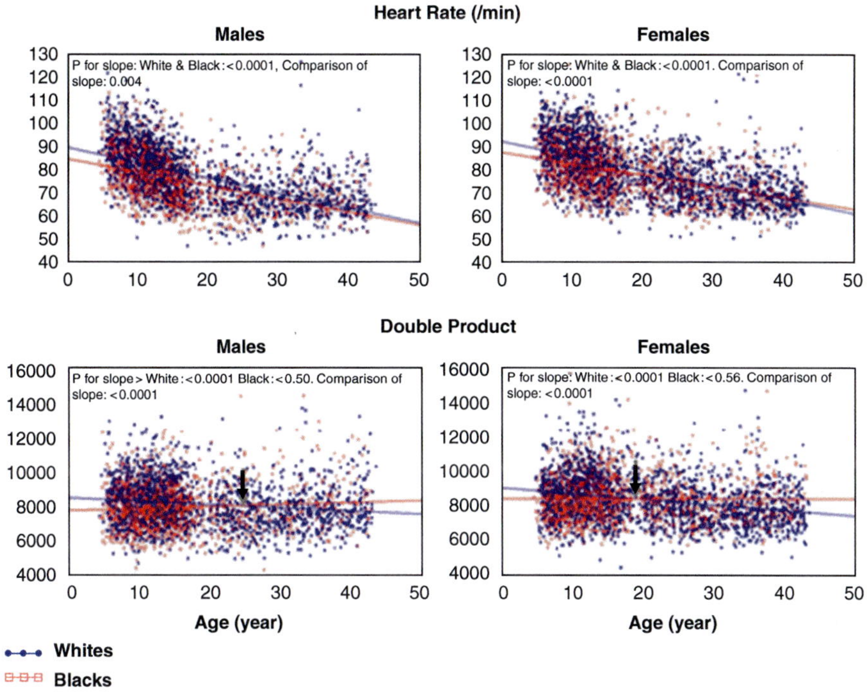

Fig. 6.5 Observations of heart rate show rates decrease with age from childhood to middle age with greater levels in white individuals while heart rate x systolic BP (double product) was greater in white children and greater in black adults with a "crossover" around 25 years of age. [35]

hood to adulthood showed an increase in BP in both races, but a greater increase occurred in blacks, as is well known. Of interest, however, studies of heart rate and rate-blood pressure product (double product) produced a cross over phenomenon of being greater in white children (more from heart rate being greater in white children and blood pressure levels being about the same level in childhood) to a greater double product in black adults (more from increasing BP levels in blacks) [35]. The cross-over occurred around 25 years of age (Fig. 6.5). The increasing obesity and factors related to increasing cardiometabolic syndrome are suggested as a potential explanation of the cross-over phenomenon.

6.4 Blood Pressure Variability Studies

BP is a highly variable quantitative trait that has two aspects: (1) levels at a point in time among individuals (inter-individual variability) and (2) fluctuations within the same individual over time (intra-individual variability). Long term and multiple

Table 6.2 Variability measures of blood pressure from childhood to adulthood

	White		Black		Race difference	
	Male n=317	Female n=401	Male n=125	Female n=210	Male	Female
SBP (mmHg)						
Variability I	13.1	8.1**	17.0	13.2**	<0.001	<0.001
Variability II	4.2	3.9**	4.7	4.5	<0.001	<0.001
Variability III	7.6	6.1**	10.3	8.5**	<0.001	<0.001
DBP (mmHg)						
Variability I	11.7	8.7**	12.4	10.9**	0.100	<0.001
Variability II	4.0	4.0	4.4	4.5	0.017	<0.001
Variability III	7.5	5.9**	8.9	7.4**	<0.001	<0.001

Variability I = age-related trend; Variability II = deviations around age-predicted values; Variability III = deviations from overall mean
Sex difference within racial group: **$p < 0.01$

examinations as obtained by serial examinations over years in the Bogalusa Heart Study allowed repeated and later observations (intra-individual variability). A relation of BP variability to changes in cardiovascular system including cardiac enlargement, left ventricular mass index g/m$^{2.7}$ on young to middle-aged adults (mean age 38.4 years), was also available for comparison. This cohort study provided 4–14 serial examinations from childhood, with an average of 19.7 years of follow up [36, 37].

Predictor variables were noted as (1) age related BP trends, (2) deviation of BP around age-predicted values and (3) deviations from mean levels of BP. These variability measures are shown in Table 6.2. Greater and more consistent changes were noted for systolic BP, as might be expected. A relation of systolic BP variability on cardiac enlargement was found significant only in blacks (Fig. 6.6) [36]. These observations indicate a greater response to stimuli and a greater variability of BP over time in black individuals. Similar findings were significant for diastolic BP. Variability effects account for more severe cardiovascular-renal changes over time producing a greater BP burden in blacks, as monitored by echo changes of cardiac enlargement. The well-known clinical evidence of more heart failure and cardiac enlargement in black individuals are being reflected in these community-based longitudinal young cohort studies.

6.5 Control Mechanisms Related to Heart Rate and Blood Pressure

Different mechanisms operating in control of BP and in control of heart rate are suggested as depicted in Table 6.3. Heart rate is controlled more by the autonomic nervous system, sympathetic/parasympathetic balance, while BP control may be even more complex. BP control occurs, not only by sympathetic/parasympathetic balance, but by endothelial derived relaxing factor, status of the vascular endothe-

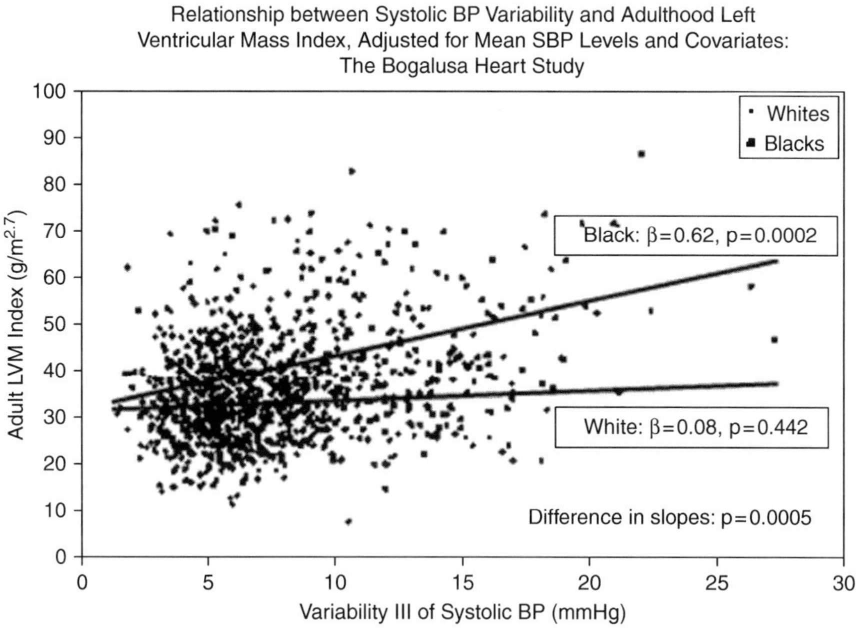

Fig. 6.6 Systolic BP variability related to left ventricular mass. Note the increasing relation in blacks. [36]

lium, renin-angiotensin-aldosterone system, and structure of the vascular wall. We were able to show the relationship of birth weight with changes of BP over time was modulated by β adrenergic receptor genes [37]. Racial contrasts noted in these studies are summarized below.

Blacks vs. whites and males vs. females show greater BP variability. Black vs. whites with greater variability showed greater LVM index, evidence of target organ changes of the cardiovascular system, vascular damage implicating endothelial integrity, nitric oxide production, activity of renin-angiotensin-aldosterone system and balance in the autonomic nervous system response. Slower resting heart rate and

Table 6.3 Different pathways that affect heart rate and blood pressure by race

Heart rate:

Autonomic nervous system

- Sympathetic—Increased heart rate through norepinephrine via β receptors
- Parasympathetic—Decreased heart rate though muscarinic receptors

Blood Pressure:

Autonomic nervous system

- Sympathetic—Increases vasoconstriction through α receptors
- Parasympathetic—Dilation through muscarinic receptors

Endothelial derived relaxing factor—NO

Renin-angiotensin-aldosterone system

- Angiotensin as marked vasoconstrictor and CV damage

heart rate variability suggest greater parasympathetic activity in blacks, but faster resting heart rate and faster response of heart rate to stimuli suggested more central sympathetic activity in whites. The greater BP response to various stimuli suggested a more reactive sympathetic response of blood pressure in blacks with stress.

6.6 Conclusion

By conjecture, these findings indicate long-term BP variations reflecting stimulus or stress responses underlie subclinical "silent" cardiovascular system changes. Heart rate control and heart rate variability studies indicate a greater parasympathetic autonomic control in blacks at rest but a greater blood pressure response occurs in blacks to various stimuli, a more reactive sympathetic activity under stress. Even low birth weight may be a precursor of such changes. These observations reflect different mechanisms controlling both heart rate and BP. Such observations are consistent with greater vascular structure/function changes found beginning at a young age in blacks, while more coronary atherosclerotic changes in whites, especially white males [38–40].Potentially, these observations have implications for both treatment and preventive cardiology.

References

1. Chobanian AV, Bakris GL, Black HR, Cushman WC, Green LA, Izzo JL Jr, Jones DW, Materson BJ, Oparil S, Wright JT Jr, Roccella EJ (2003) Joint national committee on prevention, detection, evaluation, and treatment of high blood pressure. National heart, lung, and blood institute; national high blood pressure education program coordinating committee. Seventh report of the Joint National committee on prevention, detection, evaluation, and treatment of high blood pressure. Hypertension 42:1206–1252
2. Devereux RB, James GD, Pickering TG (1993) What is normal blood pressure? Comparison of ambulatory pressure level and variability in patients with normal or abnormal left ventricular geometry. Am J Hypertens 6:211S–215S
3. Schwartz GL, Turner ST, Moore JH, Sing CF (2000) Effect of time of day on intraindividual variability in ambulatory blood pressure. Am J Hypertens 13:1203–1209
4. Rothwell PM, Howard SC, Dolan E, O'Brien E, Dobson JE, Dahlöf B, Sever PS, Poulter NR (2010) Prognostic significance of visit-to-visit variability, maximum systolic blood pressure, and episodic hypertension. Lancet 375:895–905
5. Rothwell PM (2010) Limitations of the usual blood-pressure hypothesis and importance of variability, instability, and episodic hypertension. Lancet 375:938–48 (Review)
6. Hathaway DK, D'Agostino RB (1993) A technique for summarizing longitudinal data. Stat Med 12:2169–2178
7. Grove JS, Reed DM, Yano K, Hwang LJ (1997) Variability in systolic blood pressure—a risk factor for coronary heart disease? Am J Epidemiol 145:771–776
8. Mancia G, Di Rienzo M, Parati G, Grassi G (1997) Sympathetic activity, blood pressure variability and end organ damage in hypertension. J Hum Hypertens 11 (Suppl 1):S3–S8 (Review)

9. Kikuya M, Hozawa A, Ohokubo T, Tsuji I, Michimata M, Matsubara M, Ota M, Nagai K, Araki T, Satoh H, Ito S, Hisamichi S, Imai Y (2000) Prognostic significance of blood pressure and heart rate variabilities: the ohasama study. Hypertension 36:901–906
10. Mancia G, Grassi G (2000) Mechanisms and clinical implications of blood pressure variability. J Cardiovasc Pharmacol 35(7, Suppl 4):S15–S19 (Review)
11. Mancia G (2009) Defining blood pressure goals: is it enough to manage total cardiovascular risk? J Hypertens Suppl 27:S3–S8
12. Barker DJP, Osmond C, Golding J, Kuth D, Wadsworth MEJ (1989) Growth in utero, blood pressure in childhood and adult life, and mortality from cardiovascular disease. BMJ 298:564–567
13. Nilsson PM, Östergren P-O, Nyberg P, Söderström M, Allebeck P (1997) Low birth weight is associated with elevated systolic blood pressure in adolescence: a prospective study of a birth cohort of 149 378 Swedish boys. J Hypertens 15:1627–1631
14. Leon DA, Johansson M, Rasmussen F (2000) Gestational age and growth rate of fetal mass are inversely associated with systolic blood pressure in young adults: an epidemiologic study of 165136 swedish men aged 18 years. Am J Epidemiol 152:597–604
15. Gardner DS, Bell RC, Symonds ME (2007) Fetal mechanisms that lead to later hypertension. Curr Drug Targets 8:894–905
16. Mzayek F, Hassig S, Sherwin R, Hughes J, Chen W, Srinivasan SR, Berenson GS (2007) The relationship of birth weight with the developmental trends of blood pressure from childhood through mid-adulthood. The Bogalusa Heart Study. Am J Epidemiol 166:413–420
17. Nuyt AM, Alexander BT (2009) Developmental programming and hypertension. Hypertension 18:144–152
18. Law CM, de Swiet M, Osmond C, Fayers PM, Barker DJ, Cruddas AM, Fall CH (1993) Initiation of hypertension in utero and its amplification throughout life. BMJ 306:24–27
19. Davies AA, Smith GD, May MT, Ben-Shlomo Y (2006) Association between birth weight and blood pressure is robust, amplifies with age, and may be underestimated. Hypertension 48:431–436
20. Gamborg M, Byberg L, Rasmussen F, Andersen PK, Baker JL, Bengtsson C et al (2007) NordNet Study Group. Birth weight and systolic blood pressure in adolescence and adulthood: meta-regression analysis of sex- and age-specific results from 20 nordic studies. Am J Epidemiol 166:634–645
21. Chen W, Srinivasan SR, Berenson GS (2010) Amplification of the association between birthweight and blood pressure with age: the Bogalusa Heart Study. J Hypertens 28:2046–2052
22. Chen W, Srinivasan SR, Berenson GS (2010) Low birth weight is associated with higher blood pressure variability from childhood to adulthood in a biracial (black–white) cohort: the Bogalusa Heart Study. Abstract published. Gulf central chapter american society of hypertension annual meeting
23. Mzayek F, Sherwin R, Hughes J, Hassig S, Srinivasan S, Chen W, Berenson GS (2009) The association of birth weight with arterial stiffness at mid-adulthood: the Bogalusa Heart Study. J Epidemiol Community Health 63:729–733
24. Chen W, Srinivasan SR, Li S, Berenson GS (2006) Different effects of atherogenic lipoproteins and blood pressure on arterial structure and function: the Bogalusa Heart Study. J Clin Hypertens (Greenwich) 8:323–329
25. Urbina EM, Bao W, Pickoff AS, Berenson GS (1998) Ethnic (black–white) contrasts in heart rate variability during cardiovascular reactivity testing in male adolescents with high and low blood pressure: the Bogalusa Heart Study. Am J Hypertens 11:196–202
26. Burke GL, Arcilla RA, Culpepper WS, Webber LS, Chiang YK, Berenson GS (1987) Blood pressure and echocardiographic measures in children: the Bogalusa Heart Study. Circulation 75:106–114
27. Soto LF, Kikuchi DA, Arcilla RA, Savage DD, Berenson GS (1989) Echocardiographic functions and blood pressure levels in children and young adults from a biracial population: the Bogalusa Heart Study. Am J Med Sci 297:271–279

28. Voors AW, Berenson GS, Dalferes ER, Webber LS, Shuler SE (1979) Racial differences in blood pressure control. Science 204:1091–1094
29. Berenson GS, Voors AW, Webber LS, Dalferes ER Jr, Harsha DW (1979) Racial differences of parameters associated with blood pressure levels in children—the Bogalusa Heart Study. Metabolism 28:1218–1228
30. Voors AW, Webber LS, Berenson GS (1980) Racial contrasts in cardiovascular response tests for children from a total community. Hypertension 2:686–694
31. Kim JR, Kiefe CI, Liu K, Williams OD, Jacobs DR Jr, Oberman A (1999) Heart rate and subsequent blood pressure in young adults: the CARDIA study. Hypertension 33:640–646
32. Victor RG, Leimbach WN Jr, Seals DR, Wallin BG, Mark AL (1987) Effects of the cold pressor test on muscle sympathetic nerve activity in humans. Hypertension 9:429–436
33. Calhoun DA, Mutinga ML, Wyss JM, Oparil S (1994) Muscle sympathetic nervous system activity in black and caucasian hypertensive subjects. J Hypertens 12:1291–1296
34. Berenson GS, Dalferes E Jr, Savage D, Webber LS, Bao W (1993) Ambulatory blood pressure measurements in children and young adults selected by high and low casual blood pressure levels and parental history of hypertension: the Bogalusa Heart Study. Am J Med Sci 305:374–382
35. Berenson GS, Patel DA, Wang H, Srinivasan SR, Chen W (2008) Pressure-heart rate product changes from childhood to adulthood in a biracial population—a crossover phenomenon: the Bogalusa Heart Study. J Am Soc Hypertens 2:80–87
36. Chen W, Srinivasan SR, Litao R, Berenson GS (2010) Increased left ventricular mass is associated with long term blood pressure variability beginning in childhood in black adults: the Bogalusa Heart Study. Abstract published. American college of cardiology 59th annual scientific session
37. Chen W, Srinivasan SR, Hallman DM, Berenson GS (2010) The relationship between birthweight and longitudinal changes of blood pressure is modulated by beta-adrenergic receptor genes: the Bogalusa Heart Study. J Biomed Biotechnol 2010:543514
38. Newman WP 3rd, Freedman DS, Voors AW, Gard PD, Srinivasan SR, Cresanta JL, Williamson GD, Webber LS, Berenson GS (1986) Relation of serum lipoprotein levels and systolic blood pressure to early atherosclerosis. The Bogalusa Heart Study. N Engl J Med 314:138–144
39. Berenson GS, Wattigney WA, Tracy RE, Newman WP 3rd, Srinivasan SR, Webber LS, Dalferes ER Jr, Strong JP (1992) Atherosclerosis of the aorta and coronary arteries and cardiovascular risk factors in persons aged 6 to 30 years and studied at necropsy (the Bogalusa Heart Study). Am J Cardiol 70:851–858
40. Malcom GT, Oalmann MC, Strong JP (1997) Risk factors for atherosclerosis in young subjects: the PDAY study. Pathobiological determinants of atherosclerosis in youth. Ann N Y Acad Sci 817:179–188 (Review)

Chapter 7
Obesity—Findings from the Bogalusa Heart Study

David S. Freedman and Heidi M. Blanck

Abstract Findings from the Bogalusa Heart Study have furthered the understanding of the natural history of obesity, as well as its short- and long-term consequences. Although the limitations of BMI (kg/m^2) as an indicator of body fatness are widely recognized, this simple index appears to be able to identify metabolic risk as accurately as do methods that are more expensive and laborius.

Between 1973–1974 and 1992–1994, the prevalence of obesity among school-aged children in Bogalusa increased about 3-fold, from 6 to 17%. Studies have shown that obese children are likely to have adverse levels of coronary heart disease risk factors, and are at increased risk for adult obesity, diabetes mellitus, atherosclerosis and left ventricular dilatation. These consequences will become increasingly evident as persons with childhood-onset obesity age.

Because of the long follow-up periods needed to study the relation of childhood obesity to adult disease, non-invasive techniques such as B-mode ultrasonography, electron beam tomography and Doppler echocardiography, will likely provide the most useful information on the relation of childhood obesity to disease. The frequent non-adherence of adolescents to lifestyle changes and medical treatment will complicate prevention and treatment efforts.

Keywords Obesity • BMI • Lipids • Blood pressure • Diabetes • Coronary heart disease

7.1 Secular Trends in Child Obesity

The prevalence of adult obesity, typically defined as a BMI (kg/m^2) of 30 or more, has greatly increased in the U.S. and in other countries since the 1960s. In the U.S., for example, the mean weight of adults increased by about 10 kg from 1960–1962

D. S. Freedman (✉)
Division of Nutrition, Physical Activity and Obesity, Obesity Prevention and Control Branch, Centers for Disease Control and Prevention K-26, Atlanta, GA, USA
e-mail: dxf1@cdc.gov

G. S. Berenson (ed.), *Evolution of Cardio-Metabolic Risk from Birth to Middle Age*,
DOI 10.1007/978-94-007-1451-9_7, © Springer Science+Business Media B.V. 2011

to 1999–2002, and over one-third of U.S. adults are now obese [1, 2]. Similar trends have been observed among children and adolescents, with the prevalence of obesity (defined as a BMI ≥95th percentile of the CDC growth charts [3] or a BMI ≥30 kg/m^2) having increased by about 4-fold. About 19% of 6- to 19-year-olds in the U.S. are now obese, and another 16% are overweight (BMI between the CDC 85th and 94th percentiles) [4].

One of the first reports documenting a secular increase in the weight of children was from the Bogalusa Heart Study, with the mean weight of 5- to 14-year-olds increasing by about 1 kg from 1973–1974 to 1978–1979 [5]. Subsequent analyses from Bogalusa indicated that the trends in BMI accelerated in the 1980s and 1990s, and increases were also observed in the thickness of the triceps skinfold [6]. Smoothed levels of BMI by age in 4 of the cross-sectional examinations in Bogalusa are shown in Fig. 7.1. Based on the CDC cut-points, the prevalence of obesity in Bogalusa increased from 6 (1973–1974) to 17% (1992–1994), and the combined prevalence of overweight and obesity increased from 15 to 32% during this period. The secular increases in BMI were not evenly distributed across the entire distribution, but were more pronounced at the upper end of the BMI distribution, with the 90th percentile of BMI increasing much more than the 50th percentile [6]. National data have also shown that the positive skewness of the distribution of BMI levels has increased [7].

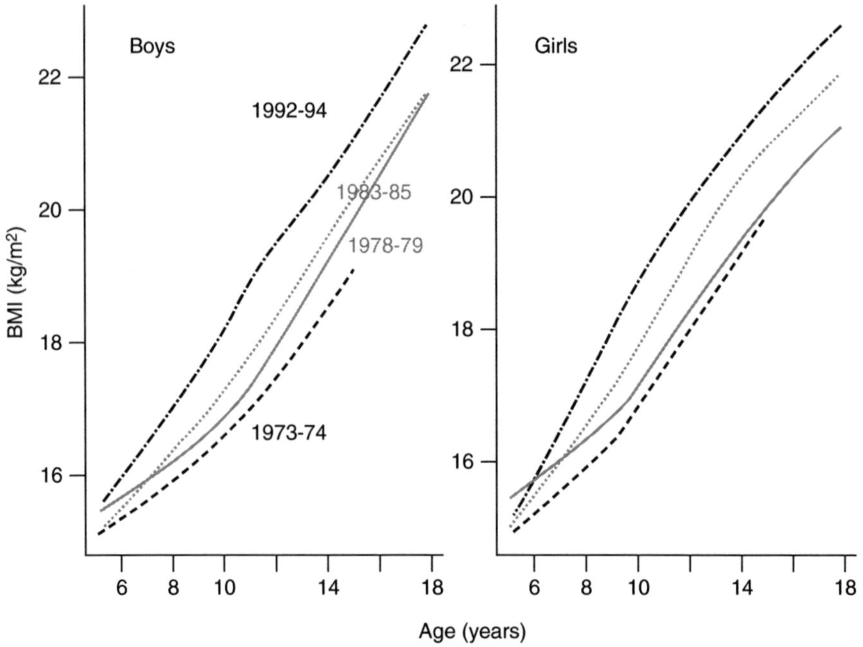

Fig. 7.1 Smoothed levels of BMI by age among children in 4 cross-sectional examinations in Bogalusa: 1973–1974, 1978–1979, 1983–1985, and 1992–1994 examinations. Curves were smoothed using lowess. [101]

Although U.S. data suggest that the secular trends have slowed or stabilized since 2000 [4], it is possible that the prevalence of obesity in still increasing among some race-ethnic groups or in some geographic regions [8]. In Bogalusa, the mean BMI of 10- to 17-year-olds increased by about 3 kg/m^2 between 1992–1994 and 2008–2009, resulting in a doubling of the prevalence of obesity [9]. Almost one third of examined 10- to 17-year-olds in Bogalusa were obese in 2008–2009, a prevalence that is about 50% higher than the overall U.S. prevalence.

7.1.1 Possible consequences of the secular trends

Data from the Bogalusa Heart Study indicate that these secular trends in BMI have been associated with higher levels of triglycerides and lower levels of HDL cholesterol over time [10]. Although some investigators [11] have reported an increase in the blood pressure of U.S. children (from 1988–1984 to 1999–2000), and have attributed this trend to increases in BMI, blood pressure levels of Bogalusa participants have not increased over time [10]. It should be realized that blood pressure levels among children are associated with height, as well as with BMI and age [12]. Because there have been secular increases in the height of children and adolescents in the U.S. over the last few decades [1], any observed changes in blood pressure levels may be due to trends in height rather than weight. It has been concluded that it is uncertain if there have been secular trends in levels of lipids and blood pressure among children [13].

Obesity is the most important risk factor for type 2 diabetes, and the results of several clinic-based studies in the 1990s suggested that the recent increases in obesity would markedly increase the prevalence of type 2 diabetes among children and adolescents [14]. For example, the annual number of newly diagnosed cases of type 2 diabetes among adolescents in the Cincinnati area increased about 10-fold (from 0.7 to 7.2 per 100,000) between 1982 and 1994 [15]. Obesity is very common among these newly diagnosed childhood cases, with mean BMIs ranging from 29 to 38 kg/m^2 across studies [16].

Although the prevalence of type 2 diabetes has risen among children, probably as a result of the secular trends in obesity, the overall prevalence of physician-diagnosed type 2 diabetes remains low. Based on population ascertainment in 6 geographic regions that cover 6% of all children and adolescents in the U.S., it was estimated that the prevalence of type 2 diabetes was about 0.3 per 10,000 among children (<10 years) and 4 per 10,000 among 10- to 19-year-olds [17]. A prevalence of 4 per 10,000 in a population of 41.5 million [18] would indicate that about 16,500 10- to 19-year-olds in the U.S. have overt type 2 diabetes.

In parallel with the race/ethnic differences seen for child obesity, type 2 diabetes is also more frequently observed among minorities than among whites, with prevalences per 10,000 (among 10- to 19-year-olds) of 17 among American Indians, 10 among blacks, 5 among Hispanics, and 2 among whites [17]. The poor glycemic control seen among adolescents with type 2 diabetes [19] and the longer duration of child-onset diabetes [20] may increase the risk of subsequent complications. Longitudinal studies from Bogalusa have shown that childhood levels of BMI and fasting glucose are predictive of pre-diabetes (fasting glucose of 100–125 mg/dL) and type 2 diabetes among adults (ages 19–44 years) [21].

7.2 Measurement and Classification of Obesity

Despite the well known limitations of BMI in assessing the body fatness of adults, this index continues to be widely used to screen for overweight and obesity. The interpretation of BMI levels among children and adolescents is further complicated by the large increases in weight and height that occur during growth and development, with the median BMI increasing by 50% between the ages of 5 and 19 years (Fig. 7.1). Although several investigators have concluded that many children with excess body fatness do not have a high BMI (low sensitivity) [22–24], these findings may be due to the use of cut-points that result in a much higher prevalence of excess body fatness than high BMI. If cut-points are defined so that the prevalences of high levels of BMI and body fatness are approximately equal, about 75% of children with excess body fatness have a high BMI [25].

Because BMI is positively associated with age and height, some investigators have concluded that its use among children and adolescents is problematic [26]. Although several reports from the Bogalusa Heart Study have used weight/height^3 rather than BMI as an index of relative weight [5, 27, 28], other analyses [29] have shown that height is positively correlated with both body fatness (assessed by skinfold thicknesses) and with fasting insulin levels among young children (Fig. 7.2). Therefore, the greater BMI of taller children appears to appropriately reflect their increased adiposity. Furthermore, levels of BMI among children are a better predictor of adult obesity than is weight/height^3 [29]. National [30] and international [31] classification systems for child obesity are now based on BMI, but it should be realized that the body fatness of children with the same sex, age and BMI can vary greatly [32, 33].

The classification of BMI levels among children and adolescents must account for differences by sex and age, and this can be achieved by expressing a child's BMI as a z-score (standard deviation score) or percentile relative to children of the same sex and age in a reference population [34]. This allows comparisons to be made between boys and girls, as well as across ages, studies and time periods. The two most widely used classification systems for overweight and obesity among children are the CDC Growth Charts, based on children examined in the U.S. between 1963–1965 and 1988–1994 [30], and the International Obesity Task Force (IOTF) BMI cut-points [31].

Although both systems are based on the distribution of BMI levels in nationally representative samples, the cut-points for overweight and obesity are somewhat arbitrary. The 85th (overweight) and 95th (obesity) percentiles of BMI are used in the CDC Growth Charts, and the IOTF classification links child BMI to levels at

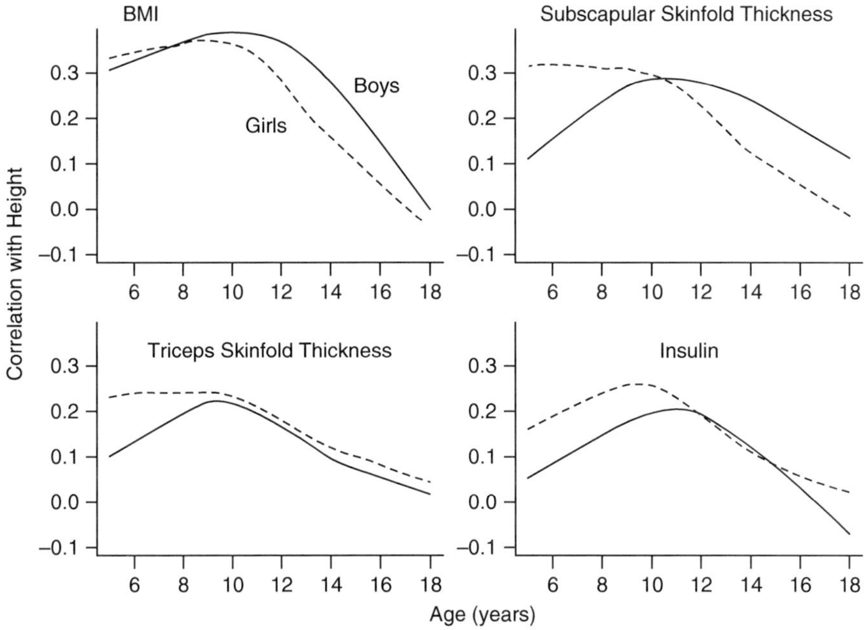

Fig. 7.2 Age differences in the magnitude of the relation of height to BMI (*upper left*), subscapular skinfold thickness (*upper right*), triceps skinfold thickness (*lower left*) and insulin (*bottom right*). Spearman correlations between height and each characteristic were calculated by sex and half year of age. Lowess was used to smooth these estimates among girls (*dashed lines*) and boys (*solid lines*). [29]

age 18 years rather than some other age in adulthood such as 25 or 30 years. In the IOTF classification, for example, 30 kg/m^2 was the 96.7th percentile among boys in the U.S. at age 18 years, and therefore the age-specific 96.7th percentile of BMI was used among younger boys (ages 2–17 years) in the U.S. as the cut-point for obesity. Because BMI levels increase with age among adults, if the cut-points had been based on the distribution of BMI levels at age 30 years (rather than at 18 years), the prevalence of adult obesity would have been greater and the BMI cut-points among children would have been substantially lower.

BMI cut-points based on associations with adverse risk factor levels have also been proposed [35–37], and the BMI cut-points in FitnessGram® [38] have been widely used [39–41]. (The FitnessGram cut-points were substantially revised in November 2010 [42, 43], and the following comparisons are based on the cut-points in use before this change). Data from Bogalusa have been used to compare the abilities of the BMI cut-points from the CDC growth charts and those from Fitnessgram to identify children who have adverse risk factor levels [44]. The magnitudes of the screening statistics between high BMI and adverse risk factors were found to be relatively low (kappa=0.25), but there were only small differences between the two classification systems. Longitudinal analyses, however, indicated that the agreement between initial and follow-up Fitnessgram classifications was substantially lower than the agreement between serial CDC classifications (kappa=0.28 vs. 0.49). It was concluded that the Fitnessgram cut-points did not identify children with adverse risk factor levels more accurately than did those from the CDC growth charts, and that the poor agreement over time complicated the interpretation of the Fitnessgram classification. Although additional analyses are needed to determine the accuracy and stability of the revised FitnessGram cut-points, it is very difficult to derive cut-points from risk-factor associations that are continuous and have no obvious break points [44].

Skinfold thickness measurements have been considered by many investigators to be attractive alternatives to BMI in the assessment of excess body fatness, and several studies have found that skinfold thicknesses are more strongly related to body fatness than is BMI [45–47]. There are, however, several limitations in the use of skinfold thicknesses, including large measurement errors [48, 49] and disagreement about the optimal sites to obtain measurements.

Body fat patterning influences the risk of type 2 diabetes and cardiovascular disease among adults [50], and abdominal obesity (typically assessed by the measurement of the waist circumference or the waist-to-height ratio (WHtR)) among children is associated with adverse risk factor levels [51, 52]. However, since fat distribution is more strongly associated with BMI than with levels of lipids and blood pressure, careful statistical analysis is needed to assess the independent influence of fat distribution. For example, some investigators have compared mean risk factor levels among children who were cross-classified by BMI and WHtR (or waist circumference) categories. This approach does not eliminate confounding by BMI and is unlikely to be optimal. Among normal-weight children (BMI<CDC 85th percentile), for example, those with a relatively high WHtR are likely to have a higher BMI than do children with low levels of both BMI and WHtR. Although WHtR has the potential to simplify the assessment of obesity-related risk, as it may be possible to use a single cut-point (e.g., 0.5) across all ages to identify high-risk children [53], additional studies are needed to verify that this ratio provides additional, independent information on obesity-related risk.

7.3 Consequences of Obesity

7.3.1 Cross-sectional Associations

Childhood obesity is associated with increased levels of several cardiovascular risk factors [54, 55] such as insulin resistance, dyslipidemia, blood pressure, C-reactive protein, and left ventricular mass [56–58]. Although these associations have been observed by numerous investigators, many results have been summarized using correlation coefficients or by presenting mean levels of various characteristics according to BMI categories across (internal) tertiles or quintiles. It can be difficult, therefore, to compare results across studies. The use of the CDC growth charts or the IOTF BMI cut-points [30, 31] will facilitate comparisons across studies.

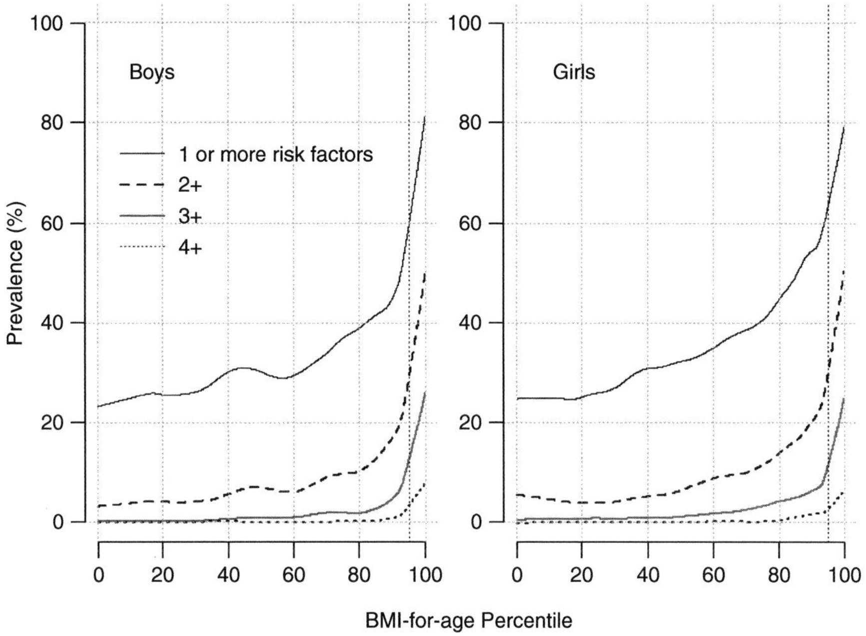

Fig. 7.3 The relation of BMI-for-age to the proportion of children with ≥1, ≥2, ≥3, and ≥4 risk factors. There were a total of 5 risk factors (high LDL-cholesterol, high triglycerides, low HDL-cholesterol, high fasting insulin, and high SBP or DBP). A high level was defined as being at or above the 90th percentile (≤10th percentile for HDL-cholesterol). Proportions were smoothed using lowess

To assess the validity of the BMI cut-points in the CDC growth charts, data from 9,167 5- to 17-year-olds in the Bogalusa Heart Study were examined [57]. Of the studied children, those who were classified as obese (prevalence, 11%) were about two times as likely (odds ratio = 2.4) as were normal-weight (BMI < 85th percentile) children to have an elevated level of total cholesterol. Odds ratios for adverse levels of other risk factors were 2.4 (DBP), 3.0 (LDL-cholesterol), 3.4 (HDL-cholesterol), 4.5 (SBP), 7.1 (triglycerides), and 12.6 (fasting insulin). In contrast, adverse risk factor levels did not vary substantially at BMI levels below the CDC 85th percentile. Although the BMI cut-points in the CDC growth charts are based on the statistical distribution of BMI levels, they perform reasonably well as a screening tool for adverse levels of lipids and blood pressure.

The relation of obesity to the clustering of multiple risk factors has also been examined in the Bogalusa Heart Study. This analysis, which classified adverse risk factors as a level ≥90th percentile (for children of the same sex and age) or ≤10th percentile (HDL-cholesterol), indicated that the relation of BMI to multiple risk factors was markedly non-linear (Fig. 7.3). For example, the prevalence of 2 or more risk factors increased from 5% (<CDC 25th percentile) to 19% (85th to 94th percentiles) to 46% (≥97th percentile) with increasing levels of BMI. Of the 333 children who had a BMI that was 120% or more of the 95th percentile (a cut-point

Table 7.1 Adjusted correlations between the adiposity measures and risk factor levels among 6,866 children and adolescents. [58]

	BMI[a]	Skinfold thickness		
		SF sum	Triceps	Subscapular
Triglycerides	0.33	0.33	0.30***	0.34
LDL cholesterol	0.19	0.19	0.17*	0.19
HDL cholesterol	−0.21	−0.20*	−0.16***	−0.19**
Fasting insulin	0.46	0.43***	0.39***	0.43***
SBP	0.28	0.24***	0.22***	0.23***
DBP	0.19	0.18	0.18*	0.18**
Risk factor summary	0.50	0.47***	0.44***	0.47***

[a] Levels of BMI, skinfold thicknesses and risk factors have been adjusted for sex, age, race, and study period. All correlation coefficients are significantly ($p < 0.001$) different from 0
P-values were calculated using the method proposed by Meng et al. [100] for comparing correlated correlation coefficients, and indicate whether the correlation between the risk factor and skinfold thickness is significantly different from the correlation between the risk factor and BMI. *$p < 0.05$, **$p < 0.01$, ***$p < 0.001$

that approximates the 99th percentile [60]), 82% had an adverse level of at least 1 risk factor and 53% had an adverse levels of at least 2 risk factors.

Analyses from Bogalusa have also compared the relative importance of BMI, skinfold thicknesses, and waist circumference in the identification of children with adverse levels of risk factors. One recent analysis [58] found that levels of lipids, fasting insulin, and blood pressures were related similarly to BMI and to the skinfold thickness sum (subscapular and triceps) (Table 7.1). For example, among children under 12 years of age, correlations with triglyceride levels were $r = 0.35$ (for both BMI and the skinfold sum), while correlations with fasting insulin were $r = 0.48$ (BMI) and $r = 0.45$ (skinfold sum). Although differences in the magnitude of the associations with BMI vs. the skinfold sum were small, almost all comparisons indicated that BMI was as strongly (or more strongly) associated with risk factor levels as was the skinfold sum. These findings confirm the results of other studies of children [46, 61] and adults [62, 63] that have found risk factor associations with BMI to be as strong as those with more accurate estimates of body fatness. Despite the limitations of BMI, it is possible that it conveys most of the relevant information on obesity-related metabolic risk.

Studies from Bogalusa have examined the additional information supplied by waist circumference or WHtR, beyond that conveyed by BMI, in identifying children with adverse risk factor levels [64, 65]. In agreement with the findings of other investigations [66, 67], it appears that BMI and WHtR have similar abilities to identify children who have adverse risk factor levels. For example, based on an overall index of multiple risk factors (lipids, fasting insulin and blood pressure), the areas under the curves were 0.85 and 0.86 ($p = 0.30$ for difference) and the multiple R^2s were 0.320 and 0.318 ($p = 0.79$) [64]. It is possible, however, that WHtR may be useful in identifying some subgroups of children who are at increased risk, such as normal-weight children with a high WHtR [68].

7.3.2 Tracking of Child Obesity into Adulthood

Children with high BMI levels are more likely to become obese adults than are thinner children [69–72], but a wide range of positive predictive values (the proportion of obese children who are obese in adulthood) and sensitivities (the proportion of obese adults who were obese in childhood) have been reported. These differences may, in part, reflect varying lengths of follow-up, ages at the initial and final examinations, the distribution of BMI levels in the sample, and the cut-points that have been used to define overweight and obesity in childhood. The influence of the recent secular trends in obesity should also be considered. For example, if subjects were first examined in the 1960s (before the large secular increases in obesity occurred) and re-examined after 2000 (after the increases), it is likely that only a small proportion of obese adults would have had a BMI in childhood that was at or above the CDC 95th percentile.

Some studies that have reported little correlation between child and adult levels of BMI [73], but this may be due to the relative thinness of the examined children (who were born in the U.K. in 1947). Differences in the BMIs of relatively thin children are likely to reflect differences in fat-free mass rather than fat mass [33], and it is likely that findings based on thin children are not applicable to the current levels of body fatness seen among U.S. children.

The strength of the association between levels of BMI in childhood and adulthood increases with childhood age [74], and positive predictive values of childhood obesity (≥CDC 95th percentile) for adult obesity ranged from 0.35 (BMI assessed at age 5 years) to 0.57 (assessed at age 15 years) in one study [75]. A much larger analysis from the Bogalusa Heart Study [76], however, found that even when childhood BMI was measured between the ages of 6 and 8 years, the prevalence of adult obesity was more than 10 times greater (78% vs. 6%) among subjects who had been obese children than among those who had a childhood BMI below the CDC 50th percentile. Further, as assessed by the magnitude of the correlation between BMI levels in childhood and adulthood, the association was only slightly weaker among subjects who were initially examined at ages of 6–8 years ($r \sim 0.6$) than for those who were first examined at ages 15–17 years ($r \sim 0.7$).

Although childhood obesity is the strongest correlate of adult obesity, other characteristics, such as parental fatness, birth weight, the timing of sexual maturation, breastfeeding, socioeconomic status and physical activity, have also been suggested as being predictive of adult BMI [70]. However, many of these characteristics are associated with child obesity, and it may be this cross-sectional association rather than the association with adult BMI, that is important. For example, several studies have shown that girls who undergo menarche at a relatively young age tend to be more obese in adulthood than are later-maturing girls [77]. However, longitudinal data from Bogalusa [78] indicate that the high prevalence of adult obesity among early maturing girls largely reflects their higher BMI levels in childhood. Controlling for childhood BMI greatly reduces the association between early menarche and adult obesity.

It has also been suggested that an early adiposity rebound increases the risk for adult obesity [79], but a young age at the BMI nadir may simply identify children whose BMI-for-age is high or is moving upwards [80]. Data from the Bogalusa Heart Study indicate that it is likely that equivalent information can be obtained more easily from a single BMI measurement at age 7 years [81]. This is likely to be easier to obtain than is a determination of the age of the BMI nadir from serial measurements.

7.3.3 Atherosclerosis and Other Indicators of Pre-clinical Disease

The initial stages of atherosclerosis begin in the aorta during childhood and adolescence, and some of these fatty streaks and fibrous plaques progress to more advanced atherosclerotic lesions. Pathology studies of children and young adults who died from external causes indicate that obesity plays a role in the development of these lesions. Among 93 decedents (ages 2–39 years) who had been previously examined in the Bogalusa Heart Study, BMI levels were found to be associated ($r=0.24$–0.41) with both fatty streaks and fibrous plaques [82]. Furthermore, these associations with BMI were comparable in magnitude to those with levels of LDL cholesterol, and were stronger than the associations with levels of DBP and HDL-cholesterol. Another study [83] of young adults (ages 15–34 years) found that BMI (at the time of death) was associated with the extent of raised lesions in the coronary arteries among men, but not among women.

The role of obesity has also been assessed using non-invasive techniques, such as electron beam tomography (EBT), that provide information on the extent of atherosclerosis in the coronary arteries [84]. Among 384 young adults (mean age, 33 years) who had been examined in the Muscatine Heart Study [85], childhood weight was found to be more predictive of EBT-determined coronary artery calcification in adulthood than were childhood levels of lipids and blood pressure.

B-mode ultrasonography is another noninvasive technique that can be used to identify persons who are at increased risk for CVD through the measurement of carotid intima-media thickness (IMT), a marker of generalized atherosclerosis. Despite the limitations of these measurements [86], carotid IMT among adults has been found to be associated with various CVD risk factors [87], arteriographically-documented lesions [88], and subsequent myocardial infarction and stroke [89]. Several investigators have found that childhood obesity is also predictive of adult IMT [90, 91], and analyses from the Bogalusa Heart Study have extended these findings.

Because obese children are likely to become obese adults [69], it is possible that the relation of childhood obesity to adult IMT reflects the importance of adult obesity rather than BMI levels in childhood [92]. Analyses from the Bogalusa Heart Study, however, indicate that the relation of child BMI to adult carotid IMT is, in part, independent of adult BMI [93]. Among 1142 participants, carotid IMT in adulthood was found to be associated with cumulative levels of BMI in both childhood and adulthood ($p<0.001$ for each association). Furthermore, the association between childhood BMI and adult IMT persisted, but was reduced ($r=0.08$), after controlling for adult BMI. Although childhood levels of lipids, lipoproteins, and blood pressures were also associated with adult IMT, these associations were not independent of adult risk factor levels.

Other studies from Bogalusa have indicated that childhood obesity is predictive of left ventricular hypertrophy [94] and left ventricular dilatation [95], and that obesity among young and middle-age adults is associated with decreased distensibility of the brachial artery [96] and decreased elasticity of the common carotid artery [97]. Furthermore, it appears that high BMI levels in childhood are as predictive (or more strongly predictive) of various adult outcomes, such as carotid intima-media thickness and type 2 diabetes, as are various combinations of risk factors such as metabolic syndrome among children [98] or the Cambridge Risk Score among adults [99]. It is possible, however, that the importance of child obesity is, at least in part, due to the tracking of high BMI levels from childhood to adulthood, and additional studies are needed to assess the independent influence of child obesity.

7.4 Conclusions

The prevalence of child obesity has greatly increased since the 1970s, and findings from the Bogalusa Heart Study have increased the understanding of the natural history of obesity, as well as it short- and long-term consequences. Obese children are at increased risk for adverse levels of CHD risk factors, diabetes mellitus, adult obesity, atherosclerosis and left ventricular dilatation. These consequences are likely to become increasingly evident due to the recent secular increases in obesity. In addition, the risks associated with the very high BMI levels currently seen among children may be substantially greater than those associated with the less severe levels of childhood overweight seen before 1970.

Because of the long follow-up periods needed to study the relation of childhood obesity to CHD, non-invasive techniques such as B-mode ultrasonography, electron beam tomography and Doppler echocardiography, will likely provide the most useful information on the relation of childhood obesity to atherosclerosis and hypertensive target organ changes. It is, however, possible that the frequent non-adherence of adolescents to lifestyle changes and medical treatment will complicate prevention and treatment efforts.

References

1. Ogden CL, Fryar CD, Carroll MD, Flegal KM (2004) Mean body weight, height, and body mass index, United States 1960–2002. Adv Data 347:1–17
2. Flegal KM, Carroll MD, Ogden CL, Curtin LR (2010) Prevalence and trends in obesity among US adults, 1999–2008. J Am Med Assoc 303:235–241

3. Kuczmarski RJ, Ogden CL, Grummer-Strawn LM, Flegal KM, Guo SS, Wei R et al (2000) CDC growth charts: United States. Adv Data 314:1–27
4. Ogden CL, Carroll MD, Curtin LR, Lamb MM, Flegal KM (2010) Prevalence of high body mass index in US children and adolescents, 2007–2008. J Am Med Assoc 303:242–249
5. Shear CL, Freedman DS, Burke GL, Harsha DW, Webber LS, Berenson GS (1988) Secular trends of obesity in early life: the Bogalusa Heart Study. Am J Public Health 78:75–77
6. Freedman DS, Srinivasan SR, Valdez RA, Williamson DF, Berenson GS (1997) Secular increases in relative weight and adiposity among children over two decades: the Bogalusa Heart Study. Pediatrics 99:420–426
7. Flegal KM, Troiano RP (2000) Changes in the distribution of body mass index of adults and children in the US population. Int J Obes Relat Metab Disord 24:807–818
8. Madsen KA, Weedn AE, Crawford PB (2010) Disparities in peaks, plateaus, and declines in prevalence of high BMI among adolescents. Pediatrics 126:434–442
9. Broyles S, Katzmarzyk PT, Srinivasan SR, Chen W, Bouchard C, Freedman DS et al (2010) The pediatric obesity epidemic continues unabated in Bogalusa, Louisiana. Pediatrics 125:900–905
10. Gidding SS, Bao W, Srinivasan SR, Berenson GS (1995) Effects of secular trends in obesity on coronary risk factors in children: the Bogalusa Heart Study. J Pediatr 127:868–874
11. Muntner P, He J, Cutler JA, Wildman RP, Whelton PK (2004) Trends in blood pressure among children and adolescents. J Am Med Assoc 291:2107–2113
12. Berenson GS, Voors AW, Webber LS, Frerichs RR (1978) Blood pressure in children and its interpretation. Pediatrics 61:333–336
13. Eisenmann JC (2003) Secular trends in variables associated with the metabolic syndrome of North American children and adolescents: a review and synthesis. Am J Hum Biol 15:786–794
14. Jackson-Leach R, Lobstein T (2006) Estimated burden of paediatric obesity and co-morbidities in Europe. Part 1. The increase in the prevalence of child obesity in Europe is itself increasing. Int J Pediatr Obes 1:26–32
15. Pinhas-Hamiel O, Dolan LM, Daniels SR, Standiford D, Khoury PR, Zeitler P (1996) Increased incidence of non-insulin-dependent diabetes mellitus among adolescents. J Pediatr 128:608–615
16. Dabelea D, Pettitt DJ, Jones KL, Arslanian SA (1999) Type 2 diabetes mellitus in minority children and adolescents. An emerging problem. Endocrinol Metab Clin North Am 28:709–729
17. SEARCH for Diabetes in Youth Study Group (2006) The burden of diabetes mellitus among US youth: prevalence estimates from the SEARCH for Diabetes in Youth Study. Pediatrics 118:1510–1518
18. U.S. Census Bureau. General demographic characteristics: 2008 population estimates. http://factfinder.census.gov/servlet/QTTable?-ds_name=PEP_2008_EST&-qr_name=PEP_2008_EST_DP1&-geo_id=01000US. Accessed 21 April 2011
19. Petitti DB, Klingensmith GJ, Bell RA, Andrews JS, Dabelea D, Imperatore G et al (2009) Glycemic control in youth with diabetes: the SEARCH for diabetes in Youth Study. J Pediatr 155:668–672
20. Lee JM (2008) Why young adults hold the key to assessing the obesity epidemic in children. Arch Pediatr Adolesc Med 162:682–687
21. Nguyen QM, Srinivasan SR, Xu JH, Chen W, Berenson GS (2010) Fasting plasma glucose levels within the normoglycemic range in childhood as a predictor of prediabetes and type 2 diabetes in adulthood: the Bogalusa Heart Study. Arch Pediatr Adolesc Med 164:124–128
22. Reilly JJ, Dorosty AR, Emmett PM (2000) Identification of the obese child: adequacy of the body mass index for clinical practice and epidemiology. Int J Obes Relat Metab Disord 24:1623–1627
23. Zimmermann MB, Gubeli C, Puntener C, Molinari L (2004) Detection of overweight and obesity in a national sample of 6–12-y-old Swiss children: accuracy and validity of reference values for body mass index from the US Centers for disease control and prevention and the International Obesity Task Force. Am J Clin Nutr 79:838–843

24. Neovius M, Rasmussen F (2008) Evaluation of BMI-based classification of adolescent overweight and obesity: choice of percentage body fat cutoffs exerts a large influence. The COMPASS study. Eur J Clin Nutr 62:1201–1207
25. Freedman DS, Wang J, Thornton JC, Mei Z, Sopher AB, Pierson RN Jr et al (2009) Classification of body fatness by body mass index-for-age categories among children. Arch Pediatr Adolesc Med 163:805–811
26. Franklin MF (1999) Comparison of weight and height relations in boys from 4 countries. Am J Clin Nutr 70:157S–162S
27. Freedman DS, Burke GL, Harsha DW, Srinivasan SR, Cresanta JL, Webber LS et al (1985) Relationship of changes in obesity to serum lipid and lipoprotein changes in childhood and adolescence. J Am Med Assoc 254:515–520
28. Newman WP III, Freedman DS, Voors AW, Gard PD, Srinivasan SR, Cresanta JL et al (1986) Relation of serum lipoprotein levels and systolic blood pressure to early atherosclerosis. The Bogalusa Heart Study. N Engl J Med 314:138–144
29. Freedman DS, Khan LK, Serdula MK, Dietz WH, Srinivasan SR, Berenson GS (2004) Interrelationships among childhood BMI, childhood height, and adult obesity: the Bogalusa Heart Study. Int J Obes Relat Metab Disord 28:10–16
30. Kuczmarski RJ, Ogden CL, Guo SS, Grummer-Strawn LM, Flegal KM, Mei Z et al (2002) 2000 CDC Growth charts for the United States: methods and development. Vital Health Stat 11(246):1–190
31. Cole TJ, Bellizzi MC, Flegal KM, Dietz WH (2000) Establishing a standard definition for child overweight and obesity worldwide: international survey. Br Med J 320:1240–1243
32. Wells JC (2000) A Hattori chart analysis of body mass index in infants and children. Int J Obes Relat Metab Disord 24:325–329
33. Freedman DS, Wang J, Maynard LM, Thornton JC, Mei Z, Pierson RN et al (2005) Relation of BMI to fat and fat-free mass among children and adolescents. Int J Obes (Lond) 29:1–8
34. Cole TJ (1990) The LMS method for constructing normalized growth standards. Eur J Clin Nutr 44:45–60
35. Dwyer T, Blizzard CL (1996) Defining obesity in children by biological endpoint rather than population distribution. Int J Obes Relat Metab Disord 20:472–480
36. Lohman TG, Going SB (2006) Body composition assessment for development of an international growth standard for preadolescent and adolescent children. Food Nutr Bull 27:S314–S325
37. Higgins PB, Gower BA, Hunter GR, Goran MI (2001) Defining health-related obesity in prepubertal children. Obes Res 9:233–240
38. Meredith MD, Welk GJ (2007) Fitnessgram activitygram. Test administration manual, 4th edn. The Cooper Institute, Human Kinetics, Champaign
39. California Department of Education (2009) 2008–2009 California physical fitness report. Summary of results. http://data1.cde.ca.gov/dataquest/PhysFitness/PFTestSt2007.asp?cYear=2008–2009&cChoice=PFTest1&RptNumber=0. Accessed 20 Aug 2010
40. New York City Department of Education. Standards/Curriculum. NYC FITNESSGRAM Assessment. http://schools.nyc.gov/Academics/FitnessandHealth/StandardsCurriculum/NYC-FITNESSGRAM.htm. Accessed 1 April 2010
41. Texas Fitnessgram. Frequently asked questions about Fitnessgram in Texas for School Year 2008–2009. http://www.fitnessgram.net/texas/txfaq/. Accessed 1 Nov 2010
42. FitnessGram®. Healthy fitness zone standards frequently asked questions. Why have new standards been developed? http://www.fitnessgram.net/newstandards/#whynew. Accessed 21 April 2011
43. FitnessGram®. The Cooper institute. Standards for healthy fitness zone, Revisions 8.6 and 9.x. www.cooperinstitute.org/ourkidshealth/fitnessgram/documents/Standardsv9final.pdf. Accessed 21 April 2011
44. Freedman DS, Fulton JE, Dietz WH, Pan L, Nihiser AJ, Srinivasan SR et al (2010) The identification of children with adverse risk factor levels by body mass index cutoffs from 2 classification systems: the Bogalusa Heart Study. Am J Clin Nutr 92:1298–1305

45. Bray GA, DeLany JP, Volaufova J, Harsha DW, Champagne C (2002) Prediction of body fat in 12-y-old African American and white children: evaluation of methods. Am J Clin Nutr 76:980–990
46. Steinberger J, Jacobs DR, Raatz S, Moran A, Hong CP, Sinaiko AR (2005) Comparison of body fatness measurements by BMI and skinfolds vs dual energy X-ray absorptiometry and their relation to cardiovascular risk factors in adolescents. Int J Obes (Lond) 29:1346–1352
47. Freedman DS, Wang J, Ogden CL, Thornton JC, Mei Z, Pierson RN et al (2007) The prediction of body fatness by BMI and skinfold thicknesses among children and adolescents. Ann Hum Biol 34:183–194
48. Ulijaszek SJ, Kerr DA (1999) Anthropometric measurement error and the assessment of nutritional status. Br J Nutr 82:165–177
49. WHO Multicentre Growth Reference Study Group (2006) Reliability of anthropometric measurements in the WHO multicentre growth reference study. Acta Paediatr Suppl 450:38–46
50. Wildman RP, Muntner P, Reynolds K, McGinn AP, Rajpathak S, Wylie-Rosett J et al (2008) The obese without cardiometabolic risk factor clustering and the normal weight with cardiometabolic risk factor clustering: prevalence and correlates of 2 phenotypes among the US population (NHANES 1999–2004). Arch Intern Med 168:1617–1624
51. Freedman DS, Srinivasan SR, Burke GL, Shear CL, Smoak CG, Harsha DW et al (1987) Relation of body fat distribution to hyperinsulinemia in children and adolescents: the Bogalusa Heart Study. Am J Clin Nutr 46:403–410
52. Kahn HS, Imperatore G, Cheng YJ (2005) A population-based comparison of BMI percentiles and waist-to-height ratio for identifying cardiovascular risk in youth. J Pediatr 146:482–488
53. Browning LM, Hsieh SD, Ashwell M (2010) A systematic review of waist-to-height ratio as a screening tool for the prediction of cardiovascular disease and diabetes: 0.5 could be a suitable global boundary value. Nutr Res Rev 23:1–23
54. Daniels SR, Arnett DK, Eckel RH, Gidding SS, Hayman LL, Kumanyika S et al (2005) Overweight in children and adolescents: pathophysiology, consequences, prevention, and treatment. Circulation 111:1999–2012
55. Wildman RP (2009) Healthy obesity. Curr Opin Clin Nutr Metab Care 12:438–443
56. Urbina EM, Gidding SS, Bao W, Pickoff AS, Berdusis K, Berenson GS (1995) Effect of body size, ponderosity, and blood pressure on left ventricular growth in children and young adults in the Bogalusa Heart Study. Circulation 91:2400–2406
57. Freedman DS, Dietz WH, Srinivasan SR, Berenson GS (1999) The relation of overweight to cardiovascular risk factors among children and adolescents: the Bogalusa Heart Study. Pediatrics 103:1175–1182
58. Freedman DS, Katzmarzyk PT, Dietz WH, Srinivasan SR, Berenson GS (2009) Relation of body mass index and skinfold thicknesses to cardiovascular disease risk factors in children: the Bogalusa Heart Study. Am J Clin Nutr 90:210–216
59. Freedman DS, Mei Z, Srinivasan SR, Berenson GS, Dietz WH (2007) Cardiovascular risk factors and excess adiposity among overweight children and adolescents: the Bogalusa Heart Study. J Pediatr 150:12–17
60. Flegal KM, Wei R, Ogden CL, Freedman DS, Johnson CL, Curtin LR (2009) Characterizing extreme values of body mass index-for-age by using the 2000 centers for disease control and prevention growth charts. Am J Clin Nutr 90:1314–1320
61. Geiss HC, Parhofer KG, Schwandt P (2001) Parameters of childhood obesity and their relationship to cardiovascular risk factors in healthy prepubescent children. Int J Obes Relat Metab Disord 25:830–837
62. Willett K, Jiang R, Lenart E, Spiegelman D, Willett W (2006) Comparison of bioelectrical impedance and BMI in predicting obesity-related medical conditions. Obesity (Silver Spring) 14:480–490
63. Hemmingsson E, Udden J, Neovius M (2009) No apparent progress in bioelectrical impedance accuracy: validation against metabolic risk and DXA. Obesity (Silver Spring) 17:183–187
64. Freedman DS, Kahn HS, Mei Z, Grummer-Strawn LM, Dietz WH, Srinivasan SR et al (2007) Relation of body mass index and waist-to-height ratio to cardiovascular disease risk factors in children and adolescents: the Bogalusa Heart Study. Am J Clin Nutr 86:33–40

65. Katzmarzyk PT, Srinivasan SR, Chen W, Malina RM, Bouchard C, Berenson GS (2004) Body mass index, waist circumference, and clustering of cardiovascular disease risk factors in a biracial sample of children and adolescents. Pediatrics 114:e198–e205
66. Moreno LA, Pineda I, Rodriguez G, Fleta J, Sarria A, Bueno M (2002) Waist circumference for the screening of the metabolic syndrome in children. Acta Paediatr 91:1307–1312
67. Sarria A, Moreno LA, Garcia-Llop LA, Fleta J, Morellon MP, Bueno M (2001) Body mass index, triceps skinfold and waist circumference in screening for adiposity in male children and adolescents. Acta Paediatr 90:387–392
68. Mokha JS, Srinivasan SR, Dasmahapatra P, Fernandez C, Chen W, Xu J et al (2010) Utility of waist-to-height ratio in assessing the status of central obesity and related cardiometabolic risk profile among normal weight and overweight/obese children: the Bogalusa Heart Study. BMC Pediatr 10:73
69. Power C, Lake JK, Cole TJ (1997) Measurement and long-term health risks of child and adolescent fatness. Int J Obes Relat Metab Disord 21:507–526
70. Parsons TJ, Power C, Logan S, Summerbell CD (1999) Childhood predictors of adult obesity: a systematic review. Int J Obes Relat Metab Disord 23(8):S1–S107
71. Reilly JJ, Methven E, McDowell ZC, Hacking B, Alexander D, Stewart L et al (2003) Health consequences of obesity. Arch Dis Child 88:748–752
72. Singh AS, Mulder C, Twisk JW, van Mechelen W, Chinapaw MJ (2008) Tracking of childhood overweight into adulthood: a systematic review of the literature. Obes Rev 9:474–488
73. Wright CM, Parker L, Lamont D, Craft AW (2001) Implications of childhood obesity for adult health: findings from thousand families cohort study. Br Med J 323:1280–1284
74. Whitlock EP, Williams SB, Gold R, Smith PR, Shipman SA (2005) Screening and interventions for childhood overweight: a summary of evidence for the US Preventive Services Task Force. Pediatrics 116:e125–e144
75. Guo SS, Wu W, Chumlea WC, Roche AF (2002) Predicting overweight and obesity in adulthood from body mass index values in childhood and adolescence. Am J Clin Nutr 76:653–658
76. Freedman DS, Khan LK, Serdula MK, Dietz WH, Srinivasan SR, Berenson GS (2005) The relation of childhood BMI to adult adiposity: the Bogalusa Heart Study. Pediatrics 115:22–27
77. Garn SM, LaVelle M, Rosenberg KR, Hawthorne VM (1986) Maturational timing as a factor in female fatness and obesity. Am J Clin Nutr 43:879–883
78. Freedman DS, Khan LK, Serdula MK, Dietz WH, Srinivasan SR, Berenson GS (2003) The relation of menarcheal age to obesity in childhood and adulthood: the Bogalusa Heart Study. BMC Pediatr 3:3
79. Rolland-Cachera MF, Deheeger M, Bellisle F, Sempe M, Guilloud-Bataille M, Patois E (1984) Adiposity rebound in children: a simple indicator for predicting obesity. Am J Clin Nutr 39:129–135
80. Cole TJ (2004) Children grow and horses race: is the adiposity rebound a critical period for later obesity? BMC Pediatr 4:6
81. Freedman DS, Kettel Khan L, Serdula MK, Srinivasan SR, Berenson GS (2001) BMI rebound, childhood height and obesity among adults: the Bogalusa Heart Study. Int J Obes Relat Metab Disord 25:543–549
82. Berenson GS, Srinivasan SR, Bao W, Newman WP III, Tracy RE, Wattigney WA (1998) Association between multiple cardiovascular risk factors and atherosclerosis in children and young adults. The Bogalusa Heart Study. N Engl J Med 338:1650–1656
83. McGill HC Jr, McMahan CA, Zieske AW, Sloop GD, Walcott JV, Troxclair DA et al (2000) Associations of coronary heart disease risk factors with the intermediate lesion of atherosclerosis in youth. The Pathobiological Determinants of Atherosclerosis in Youth (PDAY) Research Group. Arterioscler Thromb Vasc Biol 20:1998–2004
84. O'Rourke RA, Brundage BH, Froelicher VF, Greenland P, Grundy SM, Hachamovitch R et al (2000) American College of Cardiology/American Heart Association Expert Consensus Document on electron-beam computed tomography for the diagnosis and prognosis of coronary artery disease. J Am Coll Cardiol 36:326–340

85. Mahoney LT, Burns TL, Stanford W, Thompson BH, Witt JD, Rost CA et al (1996) Coronary risk factors measured in childhood and young adult life are associated with coronary artery calcification in young adults: the Muscatine Study. J Am Coll Cardiol 27:277–284
86. Coll B, Feinstein SB (2008) Carotid intima-media thickness measurements: techniques and clinical relevance. Curr Atherosclerosis Rep 10:444–450
87. Espeland MA, Tang R, Terry JG, Davis DH, Mercuri M, Crouse JR III (1999) Associations of risk factors with segment-specific intimal-medial thickness of the extracranial carotid artery. Stroke 30:1047–1055
88. Hodis HN, Mack WJ, LaBree L, Selzer RH, Liu CR, Liu CH et al (1998) The role of carotid arterial intima-media thickness in predicting clinical coronary events. Ann Intern Med 128:262–269
89. Nambi V, Chambless L, Folsom AR, He M, Hu Y, Mosley T et al (2010) Carotid intima-media thickness and presence or absence of plaque improves prediction of coronary heart disease risk: the ARIC (Atherosclerosis Risk In Communities) study. J Am Coll Cardiol 55:1600–1607
90. Davis PH, Dawson JD, Riley WA, Lauer RM (2001) Carotid intimal-medial thickness is related to cardiovascular risk factors measured from childhood through middle age: the Muscatine Study. Circulation 104:2815–2819
91. Raitakari OT, Juonala M, Kahonen M, Taittonen L, Laitinen T, Maki-Torkko N et al (2003) Cardiovascular risk factors in childhood and carotid artery intima-media thickness in adulthood: the cardiovascular risk in Young Finns Study. J Am Med Assoc 290:2277–2283
92. Juonala M, Raitakari M, Viikari JSA, Raitakari OT (2006) Obesity in youth is not an independent predictor of carotid IMT in adulthood. The cardiovascular cisk in Young Finns Study. Atherosclerosis 185:388–393
93. Freedman DS, Patel DA, Srinivasan SR, Chen W, Tang R, Bond MG et al (2008) The contribution of childhood obesity to adult carotid intima-media thickness: the Bogalusa Heart Study. Int J Obes (Lond) 32:749–756
94. Toprak A, Wang H, Chen W, Paul T, Srinivasan S, Berenson G (2008) Relation of childhood risk factors to left ventricular hypertrophy (eccentric or concentric) in relatively young adulthood (from the Bogalusa Heart Study). Am J Cardiol 101:1621–1625
95. Haji SA, Ulusoy RE, Patel DA, Srinivasan SR, Chen W, Delafontaine P et al (2006) Predictors of left ventricular dilatation in young adults (from the Bogalusa Heart Study). Am J Cardiol 98:1234–1237
96. Urbina EM, Brinton TJ, Elkasabany A, Berenson GS (2002) Brachial artery distensibility and relation to cardiovascular risk factors in healthy young adults (the Bogalusa Heart Study). Am J Cardiol 89:946–951
97. Urbina EM, Srinivasan SR, Kieltyka RL, Tang R, Bond MG, Chen W et al (2004) Correlates of carotid artery stiffness in young adults: the Bogalusa Heart Study. Atherosclerosis 176:157–164
98. Magnussen CG, Koskinen J, Chen W, Thomson R, Schmidt MD, Srinivasan SR et al (2010) Pediatric metabolic syndrome predicts adulthood metabolic syndrome, subclinical atherosclerosis, and type 2 diabetes mellitus but is no better than body mass index alone: the Bogalusa Heart Study and the cardiovascular risk in Young Finns Study. Circulation 122:1604–1611
99. Thomas C, Hypponen E, Power C (2006) Type 2 diabetes mellitus in midlife estimated from the Cambridge risk score and body mass index. Arch Intern Med 166:682–688
100. Meng XL, Rosenthal R, Rubin DB (1992) Comparing correlated correlation coefficients. Psychol Bull 111:172–174
101. Cleveland WS (1993) Visualizing data. Hobart Press, Murray Hill
102. Wells JC, Haroun D, Williams JE, Wilson C, Darch T, Viner RM et al (2010) Evaluation of DXA against the four-component model of body composition in obese children and adolescents aged 5–21 years. Int J Obes (Lond) 34:649–655

Chapter 8
Morbid Obesity and Premature Death in the Young

Pronabesh DasMahapatra and Camilo Fernandez Alonso

Abstract Morbid obesity and related ectopic lipid accumulation play an important role in the pathogenesis of multi-organ damage. This autopsy report of a 25 year old Caucasian male with history of sleep apnea but non-compliant to continuous positive airway pressure illustrates the consequences of chronic fatty organ changes in a morbidly obese young male.

Fat, especially central obesity, plays an important role in the dysregulation of hemodynamic, metabolic, and inflammatory processes through mechanisms that include activation of release of free fatty acids, macrophage infiltration into adipose tissue, hepatic lipogenesis, adipose renin-angiotensin-aldosterone system, sympathetic activation, insulin resistance, oxidative and inflammatory stress. Lipid accumulation in the heart, skeletal muscle, pancreas, liver, and kidney play an important role in the pathogenesis of heart failure, potential arrhythmias and diabetes. Excess free fatty acids also impair normal cell signaling through central nervous system–acting signaling proteins (adipokines) which affect airway neuromuscular control. This coupled with mechanical compression within pharyngeal tissues and lungs lead to pharyngeal collapsibility and sleep apnea. Hypoxic status sets up the potential of arrhythmias and sudden death.

Keywords Morbid obesity • Lipotoxicity • Obstructive sleep apnea • Cardiovascular disease • Multi-organ damage

P. DasMahapatra (✉)
Department of Epidemiology, Center for Cardiovascular Health, Tulane University School of Public Health and Tropical Medicine, New Orleans, LA, USA
e-mail: pdasmaha@tulane.edu, pronabesh@yahoo.com

G. S. Berenson (ed.), *Evolution of Cardio-Metabolic Risk from Birth to Middle Age*,
DOI 10.1007/978-94-007-1451-9_8, © Springer Science+Business Media B.V. 2011

8.1 Consequences of the Epidemic of Obesity

Obesity has reached epidemic proportions, especially in the U.S. and other industrialized countries. This has created devastating increase in morbidity and mortality from associated conditions like hypertension, type 2 diabetes mellitus, dyslipidemia, metabolic syndrome, obstructive sleep apnea (OSA), cerebrovascular and cardiovascular diseases, which pose tremendous challenges to our health care system [1]. Moreover, secular trend in obesity have shown alarming increase over the last few decades [2–4]. In our study population in Bogalusa, LA, the proportion of children and adolescents aged 5 to 17 years who are overweight (overweight plus obese) has more than tripled, from 14.2% in 1973–1974 to 48.4% in 2008–2009 [5]. Similarly, the proportion of obese children and adolescents has increased more than fivefold from 5.6% in 1973–1974 to 30.8% in 2008–2009 [5].

Morbid obesity is a major cause of premature death due to related ectopic lipid accumulation in the heart, pancreas, liver, skeletal muscle and kidneys. This chapter reviews chronic fatty change induced premature death of a morbidly obese young male in Bogalusa, to illustrate the grave consequences of the epidemic of obesity.

8.2 Autopsy Findings of the Case

A 25 year old Caucasian male with history of sleep apnea but non-compliant to continuous positive airway pressure (CPAP) was found dead by relatives in the morning. He was previously enrolled in the Bogalusa Heart Study and screened at 6 and 7 years of age. Despite having no anthropometric, hemodynamic or metabolic abnormalities in childhood (Table 8.1), at death the individual had excessive body mass index (BMI) of 36.2 kg/m^2. Long term studies in Bogalusa show while more in the category of obese children become obese adults (approx 75%), about 18% of normal weight children also become obese adults [6].

Table 8.1 Anthropometric and hemodynamic measures in childhood

Screening date: July 21, 1987	Screening date: May 12, 1988
• Age: 6 years	• Age: 7 years
• Weight: 22.3 kg	• Weight: 25 kg
• Height: 117.3 cm	• Height: 122.3 cm
• BMI: 16.2 kg/m^2	• BMI: 16.7 kg/m^2
• SBP: 101.33 mm Hg	• SBP: 96.7 mm Hg
• DBP 4th phase: 51.7 mm Hg	• DBP 4th phase: 48.3 mm Hg
• DBP 5th phase: 44 mm Hg	• DBP 5th phase: 24.3 mm Hg
• Pulse: 92 beats/min	• Pulse: 101 beats/min

BMI body mass index, *SBP* systolic blood pressure, *DBP* diastolic blood pressure

Table 8.2 Clinical–pathological report 7/23/2006, age–25 years

Immediate cause of death: sleep apnea
Physical examination and necropsy findings:

- Height: 6′ 2″ (1.88 m)
- Weight: 280 lbs (127.2 kg)
- BMI: 36.2 kg/m^2 (> 95th percentile)
- Heart:
  - 440 g. LV 2.2 cm, RV 0.7 cm, massive LV enlargement
  - 20% narrowing of the right coronary artery
  - 30% narrowing of the LAD
  - No calcification of coronary arteries
  - No findings compatible with infarction
- Liver: 1920 g (normal: 1400–1800 g)
  - Acute hepatic congestion with mild steatosis
- Kidneys: ~210 g each (normal: 125–170 g)
  - Acute congestion
- Pancreas: 180 g (normal: 60–100 g)
- Rest of the organs:
  - No specific abnormalities
- Toxicology: negative

LV left ventricle, *RV* right ventricle, *LAD* left anterior descending artery

Findings on autopsy summarized in Table 8.2, revealed cardiomegaly, with evidence of luminal narrowing of the right coronary and left anterior descending arteries. Cross-section of the heart revealed thickening of the right and left ventricular walls. Likewise, the liver, spleen, kidneys and pancreas were enlarged with the liver showing mild steatosis and serosanguineous fluid exudates. Pathological analysis revealed hypertensive and coronary heart disease and fatty infiltration of various organs.

Interpretation of autopsy was death due to obstructive sleep apnea (OSA), secondary to morbid obesity and associated co-morbid conditions of hypertensive and coronary heart diseases.

8.3 Patho-physiological Mechanisms and Concept of Lipotoxicity

Evidence now indicates that fat is a dynamic "endocrine organ" that tightly regulates nutritional balance by means of a complex interaction of adipocytes with their microenvironment [7, 8]. Dysfunctional adipose tissue, especially in morbidly obese, plays an important role in the dysregulation of hemodynamic, metabolic, and inflammatory processes through mechanisms that include activation of release of free fatty acids from the adipocytes, macrophage infiltration into the adipose tissue, hepatic lipogenesis, adipose renin-angiotensin-aldosterone system, sympathetic ac-

Table 8.3 Impact of obesity and related cardiovascular risk

Fat cell generated	Obesity interrelationships
• Inflammatory mediators – Cytokines—TNF-α, IL-6, CRP • Angiogenic proteins – VEGF • Metabolic regulators – Adiponectin – Leptin • Free fatty acids • Angiotensinogen • Stem cells	• Blood pressure • Serum lipids and lipoproteins • Hyperglycemia • Hyperinsulinemia • Diabetes • Cardiovascular – LVM, Cardiac dilatation – Vascular stiffness • Renal – End stage renal disease • Non-alcoholic fatty hepatitis

VEGF vascular endothelial growth factor, *LVM* left ventricular mass, *TNF* tumor necrosis factor, *IL* interleukin, *CRP* C-reactive protein

tivation, hyperleptinemia, insulin resistance, oxidative and inflammatory stress, and ectopic lipid storage [9–13]. Lipid accumulation in the heart, skeletal muscle, pancreas, liver, and kidney play an important role in the pathogenesis of heart failure, potential arrhythmias and diabetes [14]. This phenomenon known as lipotoxicity, leads to a cascade of fat cell generated changes including activation of inflammatory mediators, angiogenic proteins, metabolic regulators, which accentuate multi-organ damage [9–13, 15] (Table 8.3). In addition, excess free fatty acids also impair normal cell signaling, causing cellular dysfunction and even induce apoptotic cell death in extreme circumstances. Impaired signaling through central nervous system–acting signaling proteins (adipokines) may affect airway neuromuscular control [9–13]. This coupled with mechanical compression within pharyngeal tissues and lungs lead to pharyngeal collapsibility and OSA [15–17]. Hypoxic status sets up the potential of arrhythmias and sudden death.

8.4 Importance of Primary Prevention

The autopsy findings of this case merit consideration as it reflects the grave consequences of morbid obesity as early as the third decade of life. This underscores the importance of implementing obesity prevention programs early in life. The Bogalusa Heart Study group, have been advocating the importance of prevention beginning in childhood by introducing a comprehensive health promotion and education program beginning in kindergarten [18]. This program addresses the entire school environment—classroom, cafeteria, and physical education, role models (parents and teachers) and involves businesses, medical, and educational aspects of the community. The program has, and not only emphasizes prevention of obesity, but tobacco and alcohol use and social problems as violent behavior and drop outs. In a school recording BMI and using the President's Fitness Challenge, data provided showed that weight reduction and improved ¼ mile run times were benefits of increased physical activity, and education on nutrition [19]. This will be discussed in more details in later chapters.

References

1. Thompson D, Wolf AM (2001) The medical-care cost burden of obesity. Obes Rev 2:189–197
2. Ogden CL, Fryar CD, Carroll MD, Flegal KM (2004) Mean body weight, height, and body mass index, United States 1960–2002. Adv Data 27:1–17
3. Flegal KM, Carroll MD, Ogden CL, Curtin LR (2010) Prevalence and trends in obesity among US adults, 1999–2008. J Am Med Assoc 303:235–241
4. Ogden CL, Carroll MD, Curtin LR, Lamb MM, Flegal KM (2010) Prevalence of high body mass index in US children and adolescents, 2007–2008. J Am Med Assoc 303:242–249
5. Broyles S, Katzmarzyk PT, Srinivasan SR, Chen W, Bouchard C, Freedman DS, Berenson GS (2010) The pediatric obesity epidemic continues unabated in Bogalusa, Louisiana. Pediatrics 125:900–905
6. Freedman DS, Khan LK, Serdula MK, Dietz WH, Srinivasan SR, Berenson GS (2005) Racial differences in the tracking of childhood BMI to adulthood. Obes Res 13:928–935
7. Cusi K (2010) The role of adipose tissue and lipotoxicity in the pathogenesis of type 2 diabetes. Curr Diab Rep 10:306–315
8. Galic S, Oakhill JS, Steinberg GR (2010) Adipose tissue as an endocrine organ. Mol Cell Endocrinol 316:129–139
9. DeFronzo RA (2010) Insulin resistance, lipotoxicity, type 2 diabetes and atherosclerosis: the missing links. The Claude Bernard Lecture 2009. Diabetologia 53:1270–1287
10. Rial E, Rodríguez-Sánchez L, Gallardo-Vara E, Zaragoza P, Moyano E, González-Barroso MM (2010) Lipotoxicity, fatty acid uncoupling and mitochondrial carrier function. Biochim Biophys Acta 1797:800–806
11. Almaguel FG, Liu JW, Pacheco FJ, De Leon D, Casiano CA, De Leon M (2010) Lipotoxicity-mediated cell dysfunction and death involve lysosomal membrane permeabilization and cathepsin L activity. Brain Res 1318:133–143
12. Unger RH, Clark GO, Scherer PE, Orci L (2010) Lipid homeostasis, lipotoxicity and the metabolic syndrome. Biochim Biophys Acta 1801:209–214
13. Schaffer JE (2003) Lipotoxicity: when tissues overeat. Curr Opin Lipidol 14:281–287
14. McGavock JM, Victor RG, Unger RH, Szczepaniak LS (2006) American College of Physicians and the American Physiological Society. Adiposity of the heart, revisited. Ann Intern Med 144:517–524
15. Schwartz AR, Patil SP, Laffan AM, Polotsky V, Schneider H, Smith PL (2008) Obesity and obstructive sleep apnea: pathogenic mechanisms and therapeutic approaches. Proc Am Thorac Soc 15:185–192
16. Wolk R, Shamsuzzaman AS, Somers VK (2003) Obesity, sleep apnea, and hypertension. Hypertension 42:1067–1074
17. Carpagnano GE, Spanevello A, Sabato R, Depalo A, Palladino GP, Bergantino L, Foschino Barbaro MP (2010) Systemic and airway inflammation in sleep apnea and obesity: the role of ICAM-1 and IL-8. Transl Res 155:35–43
18. Downey AM, Frank GC, Webber LS, Harsha DW, Virgilio SJ, Franklin FA, Berenson GS (1987) Implementation of "Heart Smart:" a cardiovascular school health promotion program. J Sch Health 57:98–104
19. Berenson GS (2010) Cardiovascular health promotion for children: a model for a Parish (County)-wide program (implementation and preliminary results). Prev Cardiol 13:23–28

Chapter 9
Target Organ Damage Related to Cardiovascular Risk Factors in Youth

Elaine M. Urbina

Abstract Cardiovascular (CV) system changes are known to occur in childhood as part of the "silent" evidence of the beginning of adult heart diseases. Although autopsy studies have shown atherosclerotic accumulation of lipids and plaques in arteries of children and adolescents related to risk factors, it is now possible to study CV changes in the general population of youth. Noninvasive techniques have been developed to access underlying damage of the CV renal system. Ultrasonography measurement of carotid intima-media thickness has become an established tool for evaluating subclinical atherosclerosis in adults and is now being applied to study children, especially with hypercholesterolemia, diabetes and obesity. Other techniques are being used to study compliance and distensibility of the vascular tree, related to arterial wall stiffness resulting from hypertension and other etiologies of adult heart diseases. Noninvasive studies of the CV system are providing an excellent background to understand early changes leading to CV events in adulthood and a basis for beginning Preventive Cardiology at an early age.

Keywords Noninvasive CV changes • CV disease • Carotid intima-media thickness (cIMT) • Arterial stiffness • Vascular distensibility

9.1 Introduction

Throughout the world, cardiovascular (CV) diseases are the leading cause of death [1, 2]. The atherosclerotic process resulting in these adverse outcomes begins in childhood with autopsy studies demonstrating accumulation of lipid and plaque in the arteries of children and adolescents related to CV risk factors [3, 4]. The kidneys

E. M. Urbina (✉)
Department of Pediatrics, College of Medicine,
University of Cincinnati, Cincinnati, OH, USA
e-mail: Elaine.Urbina@cchmc.org
Preventive Cardiology, Cincinnati Children's Hospital Medical Center, Cincinnati, Ohio, USA

G. S. Berenson (ed.), *Evolution of Cardio-Metabolic Risk from Birth to Middle Age*,
DOI 10.1007/978-94-007-1451-9_9, © Springer Science+Business Media B.V. 2011

are also affected with hyalinization of renal arterioles found even at an average age at autopsy of only 19 years [5].

Non-invasive techniques have now been developed to assess damage to target organs related to CV risk factors. In this chapter we will review evidence of vascular changes in youth with elevated CV risk factors.

9.2 Vascular Target Organ Damage

9.2.1 Carotid Intima-Media Thickness

Currently, treatment of youth with elevated levels of CV risk factors is guided by cut-points derived from epidemiologic data in contrast to adult guidelines which are based on hard CV endpoints [6]. However, emerging methodology may result in the development of validated tools for non-invasive assessment of early atherosclerotic and hypertensive disease in youth. Future guidelines may then include assessment of the vasculature for risk-stratification which ultimately will guide treatment.

Measurement of carotid intima-media thickness (cIMT) using high resolution B-mode ultrasonography is now an established tool for evaluating subclinical atherosclerosis in adults [7] as it directly relates to future CV events [8–13] and is associated with CV risk factors (Fig. 9.1) [14–16]. Furthermore, the validity of cIMT

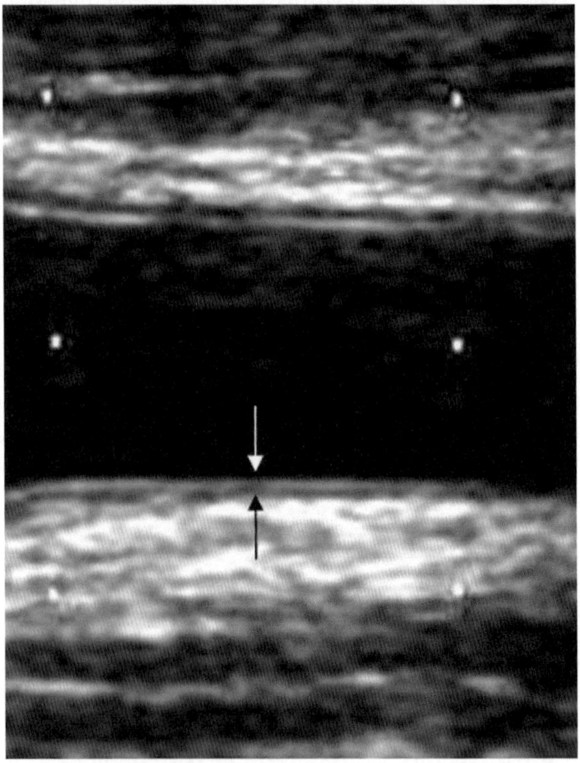

Fig. 9.1 Image taken from the distal common carotid artery demonstrating the intimal-medial complex (*between arrows*). The intimal-medial thickness is measured from the border between the echolucent vessel lumen and the echogenic intima (*white arrow*) and the border between the echolucent media and echogenic adventitia (*black arrow*). [62]

measurements has been demonstrated by direct histologic examinations proving that far wall cIMT accurately represents the intima-media thickness [17]. This technique has been proven to be sufficiently reproducible [18] to be used as an endpoint in a variety of adult intervention trials aimed at reducing CV events [18–24].

Well designed epidemiologic studies have demonstrated a strong relationship between the 'burden' of CV risk factors across childhood and future cIMT. In the Muscatine study, 725 adults age 33–42 years underwent carotid ultrasonography. Childhood total cholesterol was found to predict adult cIMT in both men and women, with increased childhood body mass index (BMI) retaining significance only for females [25]. Data from the Bogalusa Heart Study also demonstrate the adverse affect of childhood obesity on future cIMT. In this study, cIMT measured at approximately age 35 years (N=513) was compared to multiple, longitudinal (up to 6) measurements of BMI between the ages of 4 and 35 years. The correlation between childhood adiposity and cIMT was just as strong as that between current adult BMI and cIMT. Furthermore, the likelihood of having elevated cIMT was greatest among children who became obese early and maintained their high BMI (Fig. 9.2) [26]. Additional analyses on this cohort also demonstrated that higher childhood low

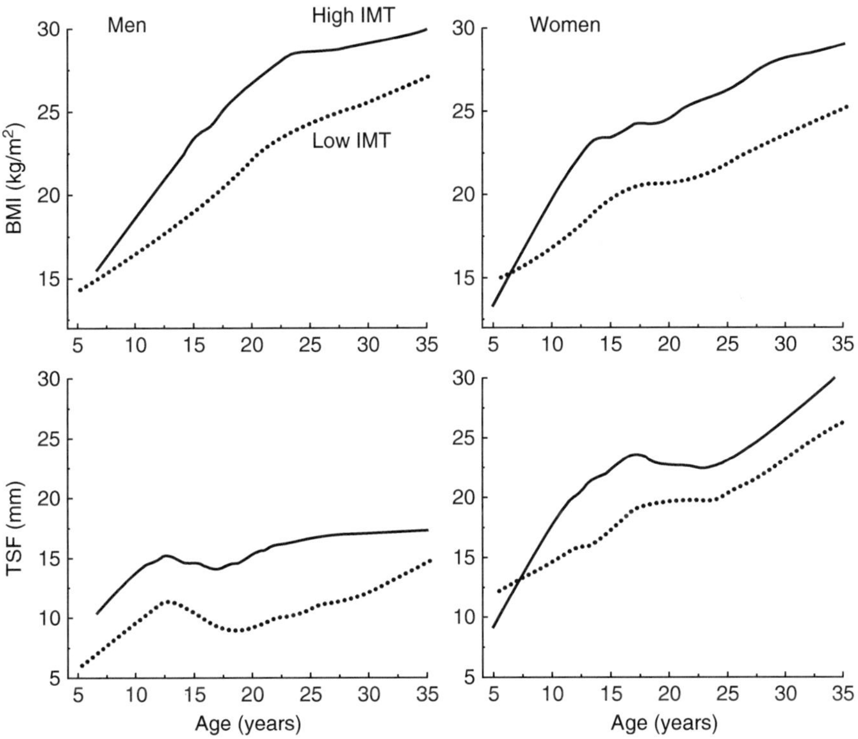

Fig. 9.2 Smoothed levels of BMI (*top panels*) and triceps skinfold thickness (TSF) (*bottom levels*) throughout life according to adult IMT; data for men are shown on the left, and for women, on the right. High IMT levels were those greater than or equal to 90th percentile, and low IMT levels were below the 10th percentile. Within each IMT group, levels of BMI and TSF were smoothed using lowess. [26]

density lipoprotein cholesterol (LDL-C) levels increased risk for thicker carotid intima as an adult (odds ratio 1.42, 95% confidence interval 1.14–1.78). Longitudinal cumulative burden of higher LDL-C (odds ratio, 1.58; 95% confidence interval, 1.24–2.01) and lower high density lipoprotein cholesterol (HDL-C) (odds ratio, 0.75; 95% confidence interval, 0.58–0.97) also significantly affected adult cIMT [27]. The Young Finns Study also related childhood CV risk factors to cIMT measured as an adult. A one standard deviation (SD) increase in adolescent BMI was associated with 0.0023 mm (95% confidence interval 1.3, 3.3) increase in common carotid cIMT in young adults [28]. This persisted after adjustment for adolescent and adult CV risk factors. Additional analyses on this cohort demonstrated that clusters of risk factors such as LDL-C, systolic blood pressure (BP), BMI and smoking at age 12–18 years was also significantly associated with greater cIMT measured at age 33–39 years even after adjusting for contemporaneous risk variable levels [29]. Unfortunately, no longitudinal evaluations have been performed comparing cIMT measured in youth to values obtained in adulthood.

Cross-sectional studies have been performed in children where assessment of cIMT is finding increasing use as a tool to evaluate vascular damage. Studies demonstrate thicker cIMT in youth with known risk factors for CV disease such as those with familial hypercholesterolemia [30–34], hypertension [35, 36], obesity [37, 39], type 1 diabetes [31, 40], and the metabolic syndrome [41]. Unfortunately, most of these studies were small (less than 100 subjects). The somewhat larger study by Sass (N=193) was able to demonstrate a relationship between BMI and systolic BP with femoral artery cIMT but not with the carotid and this relationship was true only for boys [42]. Ishizu et al. [43] also failed to demonstrate a relationship between cIMT and CV risk factors in children but this may have related to small numbers (N=60) and lack of images in other segments (bifurcation and internal carotid) as atherosclerotic disease may progress in a non-uniform fashion with individual risk factors affecting different carotid segments more strongly than others [15, 44]. Jourdan et al. studied 247 healthy children age 10–20 years. In this cohort, there was a correlation between cIMT and BMI (0.25), systolic BP (0.25) and pulse pressure (0.34, all $p \leq 0.0001$) [45]. Although providing more normative data on cIMT in the common carotid and femoral arteries, this study did not image the bifurcation or internal carotid artery segments. A recent, larger study of cIMT in children extends the observation on the relationships between CV risk factors and cIMT in youth [46]. This study examined 446 adolescents (average age 18 years) with imaging performed in the common, bifurcation and internal carotid arteries. There were significant correlations between all traditional CV risk factors and cIMT. However, the independent determinants of cIMT differed by measurement site. For the common carotid, age, gender, systolic BP and presence of type 2 diabetes mellitus (T2DM) were significant ($r^2=0.17$); for the bifurcation, age, race/ethnicity and systolic BP ($r^2=0.16$) and for the internal carotid, race/ethnicity, gender, systolic BP and total cholesterol ($r^2=0.21$). This study emphasizes the need to image all segments of the carotid artery. It also provided more data on normal cIMT in healthy youth and demonstrated the deleterious effect of obesity and obesity-related T2DM on subclinical atherosclerosis (Fig. 9.3).

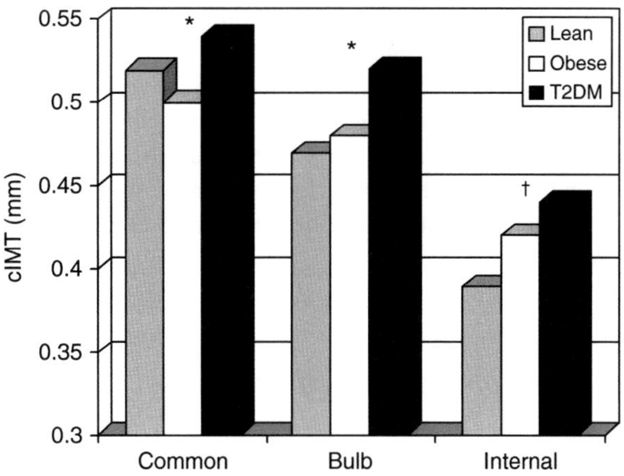

Fig. 9.3 cIMT by groups. *P<0.05, lean and obese<type 2 diabetes mellitus (T2DM); †P<0.05, lean<obese and T2DM. [46]

In younger children, measurement of IMT in the aorta (aIMT) may prove more reproducible. Studies have shown increased aIMT to be associated with low birth weight [47], intrauterine growth restriction [48], maternal smoking [49], and familial hypercholesterolemia [31]. A relationship between aIMT and seropositivity to *c pneumoniae* [50] and elevated CRP [51] was also found in children 7–11 years old, suggesting that repeated infections may accelerate vascular aging. In the Muscatine study, aIMT appeared to be more strongly related to CV risk factors than cIMT in youth under 18 years of age suggesting that measurement of aIMT may allow detection of the atherosclerotic process at an earlier age than cIMT [52].

Additional data can also be obtained by obtaining M-mode tracings of the carotid artery for measurement of maximal and minimal diameters [53]. Multiple arterial stiffness parameters can be calculated such as Pearson's and Young's elastic modulus, arterial compliance, circumferential arterial strain and beta stiffness index [54–60]. Large adult studies such as Bogalusa Heart Study have demonstrated that CV risk factors such as BP, triglycerides and BMI are independent determinants of increased common carotid artery stiffness [61].

More studies measuring cIMT in healthy children across ages, races and ethnicity are needed before cIMT can be recommended as a screening tool for early atherosclerosis in children. A recent position statement of vascular imaging techniques for use in children may stimulate additional studies in this area [62]. The promise for this technique, however, is seen in the few interventional trials in youth using cIMT as an end-point. An intensive weight loss intervention led to a significant decrease in cIMT in obese children after 1 year [63]. Wiegman et al. [64], randomized children with familial hypercholesterolemia to placebo or pravastatin and followed them for 2 years. At the end of the study, there was significant regression of cIMT in the treatment group with progression in the placebo controls [64]. This is in contrast to adult studies that do not show regression with treatment, but only maintenance of baseline carotid thickness [65]. This stresses the urgency of treating sub-clinical atherosclerosis in youth before lesions become fixed.

In conclusion, ultrasound of the carotid artery is a reliable and reproducible method for determining intima-media thickness, a non-invasive measure of atherosclerosis. This technique has been effectively applied in both young and middle-aged individuals to assess CV risk. Emerging data suggest that it will also become an important modality for CV risk stratification and to monitor the effectiveness of interventions in youth. However, more studies evaluating normal cIMT by age, race/ethnicity and sex are needed before cIMT can be incorporated into standard pediatric practice.

9.2.2 Coronary Calcification

Atherosclerotic lesions frequently become calcified with progression [66] with histologic evidence of microcrystalline calcium deposits in the lipid core of coronary plaques in subjects as young as 23 years of age (Fig. 9.4) [67]. These deposits can be imaged in the coronary arteries with electron beam computed tomography

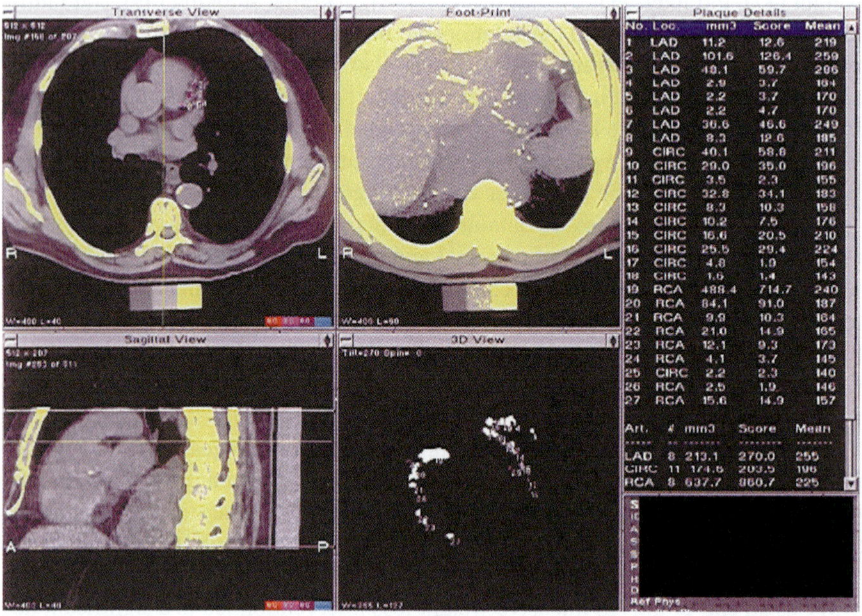

Fig. 9.4 The far right panel lists the individual lesions by vessel location, plaque volume, calcium score, and mean density (in Hounsfield units). The adjacent images illustrate several of the views that can be displayed on the workstation. From standard transverse and sagittal views (*left upper* and *lower panels*, respectively), a composite footprint image (*center upper panel*) is reconstructed and displayed. CT densities exceeding the threshold are color coded in yellow for easier identification. The lower right image shows marked coronary artery calcification in the distribution of right, anterior descending, and circumflex coronary arteries. *LAD* indicates left anterior descending coronary artery, *CIRC* circumflex artery, and *RCA* right coronary artery. [62]

(EBCT) or spiral or helical CT imaging with calcified lesions occurring almost exclusively when coronary atherosclerosis is present [68]. In adults, EBCT has proven extremely useful since it is highly sensitive in detecting lesions and defining their distribution and extent. EBCT has strong prognostic value in predicting cardiac events [69, 70]. Many studies have been performed in middle-aged and older adults documenting the prevalence of coronary artery calcification. Most demonstrate a greater prevalence with increasing age and in males in every age group paralleling the gender disparity in CV disease prevalence [69, 71–79]. EBCT also correlates with CV risk factors. In a study of 675 men and 190 women (average age 55 years), there was a significant association between self-reported history of hypertension or hypercholesterolemia and EBCT score [77]. For males, a history of diabetes, previous smoking, infrequent exercise, and obesity were also important [72]. One study found that EBCT provided additional prognostic value to the traditional Framingham risk score by identifying middle-aged women with potentially serious asymptomatic coronary artery disease despite a low score calculated from traditional CV risk factors [80].

In younger subjects, the prevalence of positive EBCT tends to be lower. One study found only 21% positive in men and 11% in females 30–39 years old [72]. The Muscatine Heart Study found a similar prevalence (31% male, 10% female) in a study of young adults in the age group 29 to 37 years [71]. Importantly, investigators demonstrated significantly increased odds for positive EBCT scan for subjects with higher BMI (odds ratio 6.4 men, 13.6 women), systolic BP (odds ratio 6.4 both genders) and low HDL-C (odds ratio 4.3 men, 4.7 women) [71]. EBCT may be even more useful as a screening tool in subjects with diabetes who have a 2 to 3 fold increased odds of having a positive EBCT scan as compared to controls when evaluated at an average age less than 40 years [81].

Few data are available describing the use of this technique in children. One study found coronary calcification in 7 out of 29 youth with heterozygous familial hypercholesterolemia [82]. Excess adiposity was also found to increase the risk for an abnormal exam. Similar studies have been conducted in youth previously diagnosed with Kawasaki disease [83]. Frey et al. [84] found that 90% (10 of 11 aneurysmic lesions) were demonstrable with ultrafast CT. A later prospective study found 95% of lesions with CT with the additional observation that the subject with the highest coronary artery calcium score was the only patient to suffer sudden death [85]. This suggests a prognostic role for EBCT in children with this specific disorder. Other pediatric patients at very high risk for adult coronary artery disease such as chronic kidney disease, may also benefit from screening with EBCT. Even when studied at an average age of only 19 years, those whose diseases were severe enough to require dialysis during childhood had a higher chance for demonstrating positive calcium scan [86]. One case study has even documented the presence of coronary calcium in two pediatric patients on dialysis [87]. Pediatric kidney [88] and heart transplant patients [89] and subjects with type 1 diabetes diagnosed in youth [90] may also be at increased risk for coronary calcification prior to reaching middle age.

In summary, limited evidence demonstrates the utility of EBCT in pediatric patients at extremely high risk for coronary artery disease. However, the low prevalence of positive scans in high risk patients under the age of 20 years and the radiation exposure associated with the test limits the utility in most pediatric settings.

9.2.3 *Arterial Stiffness*

Arterial compliance, distensibility, and stiffness measure different facets of arterial function. Compliance measures the capacity of a vessel to respond to changes in volume while distensibility is a measure of the elastic properties of an artery [91] and arterial stiffness is just the reciprocal of distensibility. Describing the functional properties of the arterial wall in vivo is difficult since distending pressure is a major determinant of all measures of arterial stiffness [92] yet the same arterial properties to be measured (vascular resistance, stiffness and wave reflections) also influence BP [93, 94]. Due to this dynamic interaction, controversy exists as to the most robust method for measurement. Furthermore, the heterogeneous composition of the arterial wall from the central arteries (high elastin to collagen ratio, decreased influence of smooth muscle tone) to the periphery (decreased ratio of elastin to collagen, increased influence of smooth muscle tone) [95, 96] mandates a variety of measurements be performed to obtain a global assessment of an individuals arterial stiffness. The types of methods most commonly used in adults include: 1) calculation of compliance or distensiblity; 2) analysis of the arterial pressure waveforms to determine the contribution of wave reflections; and 3) measuring pulse wave velocity (PWV) or time for transit of the pulse to travel from the heart to the distal site of interest.

Decreased arterial distensibility in adults is found in subjects with hypertension [97, 98], diabetes mellitus [99], and dyslipidemia [30, 98] with higher prevalence of abnormalities found with clustering of CV risk factors [61, 100]. Brachial artery distensibility was measured in 920 healthy adults (age range 18 to 38 years) as part of the Bogalusa Heart Study. Even after adjustment for age, and across a range of distending BP, decreased distensibility was seen with higher levels of CV risk factors [100]. Furthermore, subjects with clusters of metabolic syndrome risk factors had lower brachial artery distensibility [101]. Reduced distensibility may actually have prognostic significance since higher carotid stiffness at baseline actually preceded hypertension in the Atherosclerosis and Risk in Communities study [102] and predicted adverse cardiovascular outcomes in subjects with diabetes, hypertension and kidney disease [103–105].

Radial artery tonometry has also been used for calculation of large artery ('capacitive' or C1) and small vessel ('oscillatory' or C2) compliance. Both C1 and C2 were reduced in diabetic adults (average age over 65 years) undergoing angiography for coronary artery disease and this technique was useful in differentiating severity of coronary involvement [106]. In healthy younger subjects (18 to 44 years), risk factors including mean arterial pressure (MAP), BMI, insulin levels and age were inversely related to C1, accounting for 39.2% of the variance. Similarly, age, female gender and triglycerides (TG) were independent predictors of lower

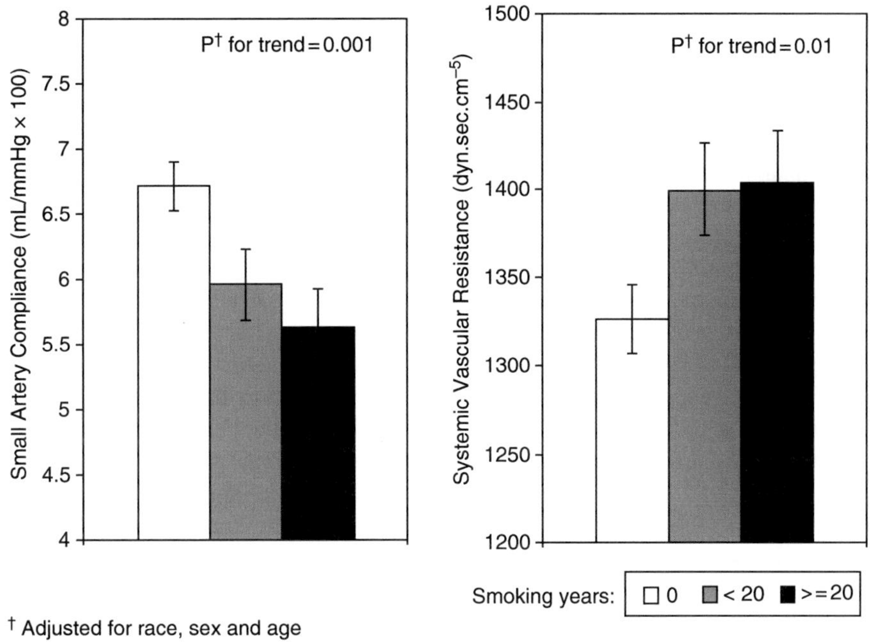

Fig. 9.5 The impact of duration of smoking on measures of pulsatile arterial function in asymptomatic, healthy young adults. The Bogalusa Heart Study. [108]

C2 [107]. This technique may be particularly useful in evaluating vascular damage from smoking as young (mean 36 years) smokers had lower C2 and higher systemic vascular resistance than non-smokers despite their lower adiposity and LDL-C (Fig. 9.5) [108]. Duration of smoking was also found to independently influence the degree of arterial dysfunction with the odds for having C2 in the lowest decile (most deterioration) 2.9 times higher in smokers (Fig. 9.5) [108]. Fortunately, the potential for improvement in arterial function with smoking cessation was also demonstrated in this cohort [109]. Although this technique has not been evaluated in children, evidence from the Minnesota Children's BP Study suggests its usefulness as an intermediate endpoint in evaluating the effect of CV risk factors measured in youth. In this study of 179 adults (average 24 years old), systolic BP was inversely related to C1 and C2 even after adjustment for gender, height, weight, insulin, HDL-C and LDL-C. They found that a 1 SD change in systolic BP was associated with a −0.30 ml/mm Hg change in C1 and a −0.008 ml/mm Hg change in C2 [110].

Early arterial wave reflections measured with tonometry have also been related to adverse CV outcomes in adults. Higher augmentation index (AIx), which reflects stiffer vessels and early wave reflection, is found in adults with coronary artery disease [111], obstructive sleep apnea [112], kidney disease [113], hypercholesterolemia [114], diabetes [115], hypertension [116] and higher levels of inflammation [117]. In fact, some investigators propose that AIx, which rises in a linear fashion with age in healthy adults, is a useful index for vascular aging [118]. Clinically, AIx is useful in evaluation of hypertensive target organ damage as higher AIx is linked to risk for left ventricular hypertrophy (LVH) [119]. The inability of beta-blockade to improve wave reflections [120] may explain the superiority of ACE inhibitors in reducing left ventricular mass (LVM) (−13.6 gm ACE vs −4.3 gm beta-blocker $p \leq 0.027$) despite similar reduction in BP [121]. Improvement in AIx may also be one explanation for the beneficial effects of a Mediterranean diet [122].

Similarly, increased PWV is strongly associated with the presence of atherosclerotic diseases (coronary heart disease, cerebrovascular disease, peripheral arterial disease, abdominal aortic aneurysm) [105]. Increased arterial stiffness measured with this technique is also associated with diabetes [123], hypertension [124], end-stage renal disease [125], hyperlipidemia [126], increasing age [127], and sedentary lifestyle [128]. In one study of hypertensive adults, subjects in the upper quartile of PWV had 7.1 times higher 95% confidence intervals (4.5 to 11.3) chance for having an elevated (>5% for 10 years) Framingham Risk Score [105]. These clustered CV risk factors promote accelerated vascular aging as demonstrated by the steeper rate of change of PWV with age in subjects with greater numbers of metabolic syndrome risk factors (Fig. 9.6) [129]. Therefore, PWV may be the most important independent predictor of cardiovascular events adding utility to risk stratification beyond that supplied by CV risk profile scores.

PWV is also emerging as a robust non-invasive measure of atherosclerotic CV disease. A strong relationship is found between CV risk factor levels measured in childhood and PWV assessment as an adult. Brachial-ankle PWV measured with an oscillometric technique was obtained in 835 subjects with at least 4 measurements of CV risk factors as part of the Bogalusa Heart Study encompassing over 25 years of follow-up. Childhood risk factors such as systolic BP and cumulative burden (area under the curve divided by follow-up years) of systolic BP, TG and smoking years since childhood were independent predictors of PWV as a young adult [130]. These findings underscore the importance of addressing CV risk factors in childhood for the prevention of accelerated vascular aging.

Although gaining acceptance as a method to improve CV risk stratification in adults, few data are available regarding pediatric usage of any of these techniques. Brachial artery distensibility using both an ultrasound and oscillometric technique

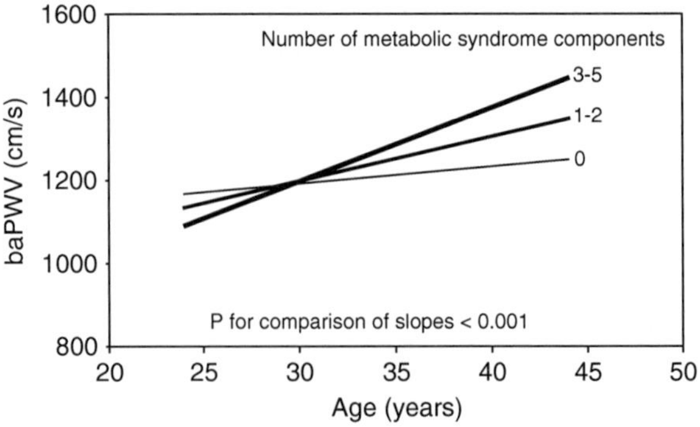

Fig. 9.6 A comparison of increasing brachial to ankle pulse wave velocity (baPWV) with age (*slope*) by the number of metabolic syndrome components, as defined by NCEP ATP III guidelines, in young adults: the Bogalusa Heart Study; P value was adjusted for race, sex, smoking, and heart rate. [129]

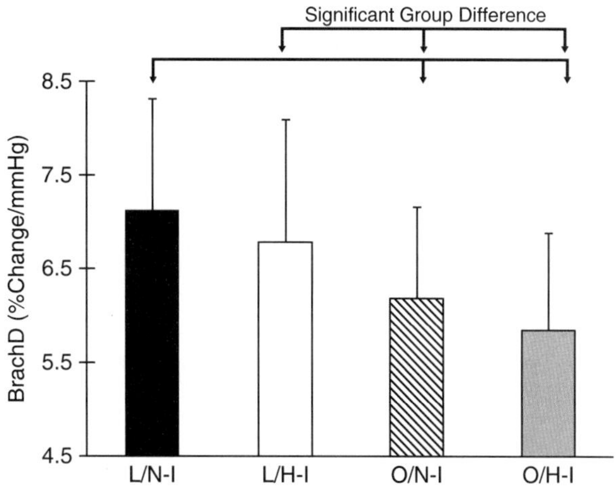

Fig. 9.7 Brachial Artery Distensibility (mean with SD bars) by BMI and Insulin groups. N=969. Significant (P<0.05) group differences indicated by arrow bars. L=Lean; O=Overweight; N-I=Normo-Insulinemic; H-I=Hyper-Insulinemic. [134]

has been evaluated in youth. Whincup et al. [131], studied 471 British children using a radio-frequency tracker device. CV risk factors including adiposity, BP, LDL-C, C-reactive protein, and insulin resistance were all negatively associated with brachial artery distensibility [131]. Using a similar technique, a significant, inverse relation was found between brachial artery distension and total cholesterol, LDL-C and apolipoprotein B [132] while other researchers found higher leptin concentrations were independently associated with impaired arterial distensibility with a −1.3% decrease seen per 10% increase in leptin (95% CI, −1.9% to −0.8%; P<0.001) [133]. In a larger study of 969 school children a linear decline in brachial distensibility was seen with the development of obesity with further deterioration in those adolescents with both obesity and hyper-insulinemia (Fig. 9.7) [134]. Further analyses indicated that obesity-related decline in adiponectin may play a role in the etiology of the vascular function decline [135]. These data suggest that measurement of brachial distensibility may be helpful in identifying early CV change in youth prior to development of the full metabolic syndrome or progression to type 2 diabetes mellitus.

Augmentation index is also proving useful in assessing vascular function in youth. Increased AIx was found in children with many conditions predisposing to CV disease such as type 1 diabetes [136, 137] and type 2 diabetes (Fig. 9.8) [138]. AIx is also higher in pediatric hemodialysis patients as compared to healthy children [139]. In young men, AIx may be helpful in differentiating true hypertension from spuriously hypertension, a condition caused by enhanced pulse wave amplification, where aortic pressure is normal but brachial BP readings are elevated [140]. Increased augmentation also differentiates adolescents and young adults with a positive family history of hypertension from unaffected families [141, 142]. Obesity also affects AIx with obese insulin-resistant youth demonstrating higher AIx than lean and obese adolescents without insulin-resistance (Fig. 9.9).

More data are available on the use of PWV in children, but most have included relatively few subjects. Increased arterial stiffness (higher PWV) was found in pediatric research subjects with type 1 diabetes mellitus [143, 144], neurofibromatosis

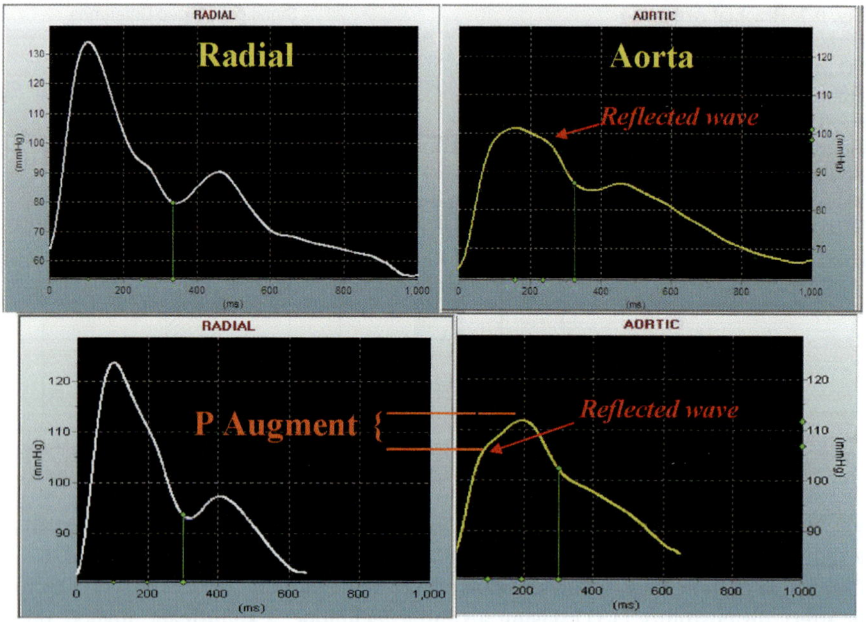

Fig. 9.8 Pulse waveform analysis obtained from radial artery tonometer demonstrating measured radial artery tracings (*left*) and calculated central aortic tracings (*right*). The adolescent with type 2 diabetes mellitus has an increased augmentation index, adjusted to a heart rate of 75 beats per minute (AIx-75) with wave reflection occurring early in systole (*bottom right*) compared with the healthy subject whose reflected wave arrived in diastole (*top right*). [62]

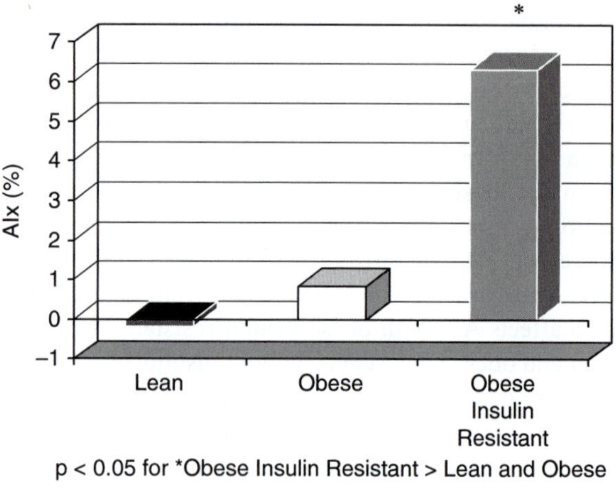

Fig. 9.9 Augmentation index is increased in youth with obesity and insulin resistance ($p \leq 0.05$, $N=343$)

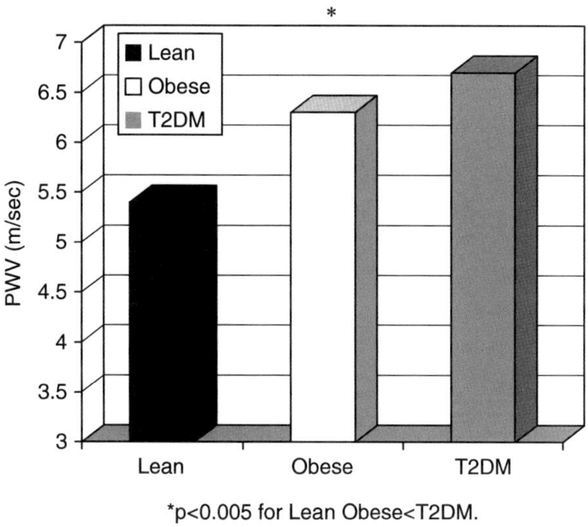

Fig. 9.10 Increased PWV found in youth with type 2 diabetes mellitus (T2DM) (N=670). [156]

[145, 146], primary snoring [147], Kawasaki disease [148], polyarteritis nodosa [149], and coarctation of the aorta after surgical repair [150–152] as compared to healthy control children. PWV may be important in evaluation of BP levels in youth since a relationship between PWV measured with an oscillometric technique and BMI and BP was demonstrated in Korean adolescents [153] and children with pre-hypertension demonstrate higher PWV than healthy youth well before a diagnosis of sustained hypertenstion is made [154]. The Georgia CV Twin Study examined heritability of PWV and found arterial stiffness to be substantially heritable with and >25% of the heritability explained by genes that also influence diastolic BP [155]. Obesity, insulin resistance and type 2 diabetes mellitus also affect PWV. A linear increase in PWV (rising stiffness) is seen when comparing healthy lean, to obese to diabetic youth (Fig. 9.10) [156]. A few published population studies have included a small number of pediatric subjects [127, 157–161]. Although insufficient data are available to set precise cutpoints for normal PWV across all pediatric age groups, race/ethnicity and sex, a recently published position paper does provide a table with PWV in normal control youth obtained with a variety of methods [62].

In summary, arterial stiffness measurements demonstrate promise in evaluation of target organ damage related to CV risk factors in youth. Higher arterial stiffness may be one mechanism for the development of left ventricular hypertrophy (LVH) in youth with CV risk factors since there is a linear relationship between global stiffness index (sum of carotid stiffness measures, brachial artery distensibilty, AIx and PWV) and LVH in youth that is independent of traditional CV risk factors (Fig. 9.11) [162]. However, the clinical use of these measures is limited by lack of validation and reproducibility studies, although there is no reason to believe that these modalities will be any less valid or reproducible in children. Certainly more data are needed to define normal cut-points for pediatric patients that account for changes in arterial size with normal growth. Studies comparing the techniques in

Fig. 9.11 Increased global stiffness index is significantly related to LVM in youth (N=670). [161]

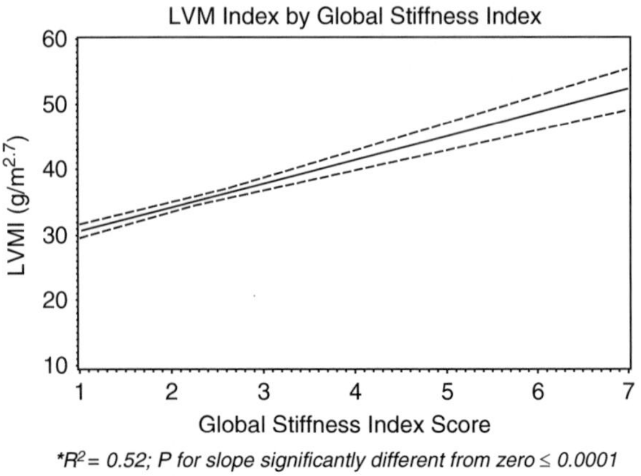

youth are also necessary as each method measures slightly different features of the arterial tree which may relate to CV risk factors with different levels of intensity.

9.2.4 *Endothelial Function*

Proper functioning of the artery depends on the endothelium's ability to balance factors promoting vasoconstriction versus vasodilation [163]. Multiple methods have been developed to assess endothelial function with flow-mediated dilation (FMD) emerging as the most well established technique used in adults. FMD reflects the ability of the endothelium to release nitric oxide in response to stress. Non-endothelial dependent dilation can also be measured after administration of a sublingual dose of nitroglycerin (NTG). This tests smooth muscle response.

In adults, both FMD and non-endothelial dependent dilation are strong predictors of adverse CV events independent of traditional CV risk factors [164] or the Framingham risk score [6]. Impaired FMD also correlates with risk factors in an asymptomatic population with strong correlations with low HDL-cholesterol [165], hypertension [165], elevated atherogenic LDL subfractions [166], insulin resistance [167], type 2 diabetes mellitus [168], tobacco exposure [169], sedentary behavior [170], and inflammation [171].

Improvement in endothelial function may be one mechanism for reduction of CV risk with treatment of obstructive sleep apnea [172, 173], increased exercise [174], and improved diet [175–182]. The CV risk-reducing effects of statins may also be accomplished not just through lipid lowering [183–186], but also through a direct effect on the endothelium [186]. The ability of anti-hypertensive therapy to improve endothelial function is less clear as neither nifedipine nor captopril treatment improved endothelial function in one study of hypertensive adults [187]. However, in adults with coronary artery disease, the ACE inhibitor quinapril, but no other medication, did result in an improvement in FMD [188].

Non-endothelial dependent dilation is less strongly related to CV risk factors although investigators have found a relationship with age, baseline diameter [165], and body mass index [189]. The fact that non-endothelial dependent dilation is not impaired in patients with familial hypercholesterolemia, essential hypertension [190] or inflammatory periodontal disease [171], suggests arterial smooth muscle function is not influenced by these disease states.

Few studies have examined non-endothelial dependent dilation measurements in children as this test requires the administration of nitroglycerin. The few data available using this technique suggest little utility in pediatric disease as there was no difference in non-endothelial dependent dilation between controls and children with familial hypercholesterolemia [191], HIV [192], renal disease [193], or after vitamin C administration in Kawasaki disease [194], although one study found oxidized LDL to be independently associated with decreased non-endothelial dependent dilation response in healthy children and those with diabetes and familial hypercholesterolemia [195].

There is, however, a growing body of literature available on FMD in the pediatric age group (Table 9.1) [62]. Decreased FMD was found in children with HIV infection [192, 196], homozygous homocystinuria [197], Kawasaki's disease [194], type 1 diabetes [198, 200], and chronic renal failure, both prior to [193] and following renal transplantation [200]. It has been hypothesized that any condition resulting in systemic inflammation in children may produce endothelial dysfunction [201, 202]. Decreases in non-endothelial dependent dilation have been observed in children with type 1 diabetes [203].

FMD also is related to CV risk factors in youth. In healthy adolescents with a positive family history of CV events, FMD is reduced, although endothelial function was even lower in children with familial hypercholesterolemia [204] or familial combined hyperlipidemia [205]. Aggoun, et al. [30] found a direct negative correlation between FMD and LDL-C but other investigators have found only lipoprotein(a) to be a significant correlate [205]. BP levels and insulin resistance are also important in determining FDM in obese children [207].

Reduced endothelial function is one mechanism through which obesity exerts its deleterious effect on the CV system. Obese children had reduced FMD compared to controls even after controlling for BP, cholesterol and glucose levels [208] with the magnitude of endothelial dysfunction correlating with BMI [208]. Severely obese children (average BMI Z-score >2 SDs above normal) had impaired FMD and non-endothelial dependent dilation which had a negative correlation with fasting insulin and apolipoprotein A-I [191]. Other CV risk factors such as second hand smoke may also influence vascular function. Non-smoking children with higher serum cotinine concentration (average age 11 years) demonstrated reduced brachial FMD compared to children with low cotinine [209]. An adverse pre-natal environment, as manifest by low birth weight was also associated with reduced FMD in adults [210], and children [211] suggesting a role for fetal programming in determination of later vascular function [212]. These data point to the need for early intervention in youth at risk for accelerated vascular aging.

Table 9.1 Conditions associated with reduced Brachial FMD. [62]

Condition	Childhood reference	Non-endothelial dependent dilation reference
Disease or Condition		
Diabetes mellitus	Jarvisalo [198], Aburawi [202]	Lee [168]*, Donaghue [203]
Dyslipidemia	De Jongh [184], Mietus-Snyder [205], Aggoun [30], Sorensen [206], Tounian [191]	Atkov [190]*, Aggoun [30]*, Jarvisalo [195]
Family history of CV disease or risk factors	Gaeta [248]	
Inflammation	Jarvisalo [201], Aburawi [202]	Amar [171]*
Age		Jensen-Urstad [189]
Adiposity or adipocytokines	Woo [208], Tounian [191], Singhal [249]	Jensen-Urstad [189], Fernandez-Real [250]
Target organ damage (LVH, increased cIMT, microalbuminuria)		Olsen [251], Papaioannou [252]
Folate or homocysteine	Wiltshire [217]	Chambers [253]*
Sedentary behavior or increased physical activity	Abbott [213]	
HIV infection	Bonnet [192], Charakida [196]	Bonnet [192]
Kawasaki's syndrome	Deng [194]	
Chronic renal disease	Kari [193], Lilien [200]	Kari [193]*
Low birth weight	Leeson [210], Leeson [211]	Leeson [210]
Intervention		
Statins	De Jongh [184]	Beckman [186]
Anti-hypertensives		Ghiadoni [187]*, Anderson [188]*
Healthy diet (Omega 3 fatty acid, Mediterranean diet, plant sterol, L-arginine, folate)	Engler [221], de Jongh [222]* Bennett-Richards [218]*, Bennett-Richards [219]	Bennett-Richards [218, 219]*
Anti-oxidants (vitamin C, Cox-2 inhibitor)	Deng [194], Mietus-Snyder [205], Engler [220]	Deng [194]* Williams [178]*, Chenevard [254]*
Exercise	Kelly [214]	

*Non-significant associations indicated with asterisk

Fortunately, studies show interventions are effective in improving endothelial function. Exercise increased FMD in healthy and obese children [63, 205, 216]. Supplements such as folate, show promise for augmenting FMD in children with type 1 diabetes [217] and chronic kidney disease [218]. However, supplementation with the nitric oxide substrate L-arginine did not improve FMD in children with chronic renal disease [219]. In children with Kawasaki disease, intravenous vitamin C injection improved brachial FMD and [194] both vitamins C and E increased FMD in children with dyslipidemia [220]. Fish oil with docosahexaenoic acid resulted in an increase in FMD after 6 weeks of therapy [221], however, plant sterols (sitosterol and campesterol) did not improve FMD in children with familial hypercholesterolemia despite their lipid lowering effects [222]. Differences in technique may account for these disparate results, but more study is certainly needed. Statin administration as in adults, demonstrate significant efficacy in improving endothelial function in children with familial hypercholesterolemia [185].

These data demonstrate the utility of FMD measurements in children with CV risk factors. However, measurement of brachial FMD and non-endothelial dependent dilation is technically challenging and involves a significant learning curve for both image acquisition and analysis [223]. There is also controversy as to which is the most valid technique for measurement and multiple sets of guidelines have been published with variations in methodology [163, 223, 224]. Furthermore, factors such as non-fasting state [225], use of vasoactive substances [223], circadian variation [226, 227], and hormones [228, 229] may influence results. Smaller children present more challenges as brachial arteries smaller than 2.5 mm in diameter are difficult to measure [223] yet are associated with a greater percent FMD [230] likely due to the greater shear stress caused by flow through a smaller vessel resulting in greater NO release [231]. Therefore, baseline size should be included as a covariate in data analyses. Time to peak dilation may also be delayed in children as compared to adults, therefore, imaging protocols should collect data out to 120 seconds [232] rather than the 60 seconds after release of occlusion recommended in adults [233].

There are also inherent difficulties in producing reproducible results, One study found the limits of agreement for repeat FMD studies were quite wide (−4.48 to 3.87%) compared to the average FMD value (11.98%) [234]. Since average FMD in controls is near 10%, variability in measurement of more than 2 to 3% will reduce ability to demonstrate differences between cases and controls [223, 235]. To complicate matters further, few data are available in healthy children and attempts to set 'normal' values has been hindered by heterogeneity in imaging technique [132, 232].

A new non-ultrasound method has been developed for assessment of endothelial function This device uses peripheral arterial tonometry (PAT) to measure blood flow in the fingertips at baseline and after an ischemic stimulus to calculate reactive hyperemia index. Reactive hyperemia index was found to be highly reproducible [236] and it correlated significantly with ultrasound-based FMD [237]. Reactive hyperemia index is reduced in patients with coronary artery disease [237] and was independently related to male sex, BMI, total to HDL-cholesterol ratio, type 2 diabetes mellitus, smoking and lipid-lowering therapy in the Framingham Heart Study [238]. This technique has not been widely evaluated in children although reduced reactive hyperemia index was found in children with congenital central hypoventilation syndrome [239], obstructive sleep apnea [240] and type 1 diabetes [241, 242]. This device does not require specialized training to operate and was found to be reproducible in children [241], however, the single finger cuff size available limits its usefulness in smaller children.

Endothelial function can be measured in younger children with doppler laser flowmetry. The advanatage of the doppler laser flowmetry technique lies in the small size of the probe which is usually applied to a hand or foot, and the variety of stimuli available to trigger an endothelial response. While young children may not tolerate minutes of ischemia induced by an inflated blood pressure cuff, the doppler laser flowmetry device can produce micro-vascular skin vasodilation with a barely perceptible heating of the probe or with iontophoresis which produces only a mild tingling when the small electrical current pulls acetylcholine or nitroprusside through the skin [243]. In adults, there is a linear decline in endothelial function measured with this technique with increasing Framingham Risk Score [244]. In children, endothelial dysfunction was found in children and adolescents with type 1 diabetes [245, 246]. Improvement was seen in doppler laser flowmetry measures with treatment of obstructive sleep apnea, a risk factor for hypertension [247]. Clearly, measurement of endothelial dysfunction in youth is necessary to determine CV risk and evaluate the usefulness of interventions. Hopefully, more pediatric data will be collected so these tests can become a routine part of pediatric primary prevention.

9.3 Conclusions

The obesity epidemic throughout the world is increasing the prevalence of elevated CV risk factors in youth. Non-invasive imaging demonstrates damage to target organs such as the heart, blood vessels and kidneys in these high risk youth. Since risk stratification scores for youth based on hard CV outcomes are not available, more data on the use of non-invasive intermediate atherosclerotic endpoints need to be collected. Improving our ability to diagnose accelerated aging related to atherosclerotic disease will allow targeting of highest risk youth decades before clinical CV disease occurs.

References

1. Kochanek KD, Smith BL (2004) Deaths: preliminary data for 2002. Natl Vital Stat Rep 52:1–47
2. Rosamond WD, Chambless LE, Folsom AR, Cooper LS, Conwill DE, Clegg L, Wang CH, Heiss G (1998) Trends in the incidence of myocardial infarction and in mortality due to coronary heart disease, 1987 to 1994. N Engl J Med 339:861–867
3. Berenson GS, Srinivasan S, Boa W, Newman WP III, Tracy RP, Wattigney WA (1998) Association between multiple cardiovascular risk factors and atherosclerosis in children and young adults. The Bogalusa Heart Study. N Engl J Med 338:1650–1656
4. A Preliminary report from the Pathological Determinants of Atherosclerosis in Youth (PDAY) Research Group (1990) A relationship of atherosclerosis in young men to serum lipoprotein cholesterol concentrations and smoking. J Am Med Assoc 264:3018–3024
5. Tracy RE, Newman WP III, Wattigney WA, Srinivasan SR, Strong JP, Berenson GS (1995) Histologic features of atherosclerosis and hypertension from autopsies of young individuals in a defined geographic population: the Bogalusa Heart Study. Atherosclerosis 116:163–179

6. Grewal J, Chan S, Frohlich J, Mancini GB (2003) Assessment of novel risk factors in patients at low risk for cardiovascular events based on Framingham risk stratification. Clin Invest Med (Medecine Clinique et Experimentale) 26:158–165
7. Greenland P, Abrams J, Aurigemma GP, Bond MG, Clark LT, Criqui MH, Crouse JR III, Friedman L, Fuster V, Herrington DM, Kuller LH, Ridker PM, Roberts WC, Stanford W, Stone N, Swan HJ, Taubert KA, Wexler L (2000) Prevention conference V: beyond secondary prevention: identifying the high-risk patient for primary prevention: noninvasive tests of atherosclerotic burden: Writing Group III. Circulation 101:E16–E22
8. Chambless LE, Heiss G, Folsom AR, Rosamond W, Szklo M, Sharrett AR, Clegg LX (1997) Association of coronary heart disease incidence with carotid arterial wall thickness and major risk factors: the Atherosclerosis Risk in Communities (ARIC) Study, 1987–1993. Am J Epidemiol 146:483–494
9. Hodis HN, Mack WJ, LaBree L, Selzer RH, Liu CR, Liu CH, Azen SP (1998) The role of carotid arterial intima-media thickness in predicting clinical coronary events. Ann Intern Med 128:262–269
10. O'Leary DH, Polak JF, Kronmal RA, Manolio TA, Burke GL, Wolfson SK Jr (1999) Carotid-artery intima and media thickness as a risk factor for myocardial infarction and stroke in older adults. Cardiovascular Health Study Collaborative Research Group. N Engl J Med 340:14–22
11. Salonen JT, Salonen R (1991) Ultrasonographically assessed carotid morphology and the risk of coronary heart disease. Arterioscler Thromb 11:1245–1249
12. Simons PC, Algra A, Bots ML, Grobbee DE, van der Graaf Y (1999) Common carotid intima-media thickness and arterial stiffness: indicators of cardiovascular risk in high-risk patients. The SMART Study (Second Manifestations of ARTerial disease). Circulation 100:951–957
13. Dijk J, van der Graaf Y, Bots M, Grobbee D, Algra A (2006) Carotid intima-media thickness and the risk of new vascular events in patients with manifest atherosclerotic disease: the SMART study. Eur Heart J 27:1971–1978
14. Grobbee D, Bots M (1994) Carotid artery intima-media thickness as an indicator of generalized atherosclerosis. J Intern Med 236:567–573
15. Urbina EM, Srinivasan SR, Tang R, Bond MG, Kieltyka L, Berenson GS (2002) Impact of multiple coronary risk factors on the intima-media thickness of different segments of carotid artery in healthy young adults (The Bogalusa Heart Study). Am J Cardiol 90:953–958
16. Selvin E, Coresh J, Golden SH, Boland LL, Brancati FL, Steffes MW (2005) Glycemic control, atherosclerosis, and risk factors for cardiovascular disease in individuals with diabetes: the atherosclerosis risk in communities study. Diabetes Care 28:1965–1973
17. Bots ML, Dijk JM, Oren A, Grobbee DE (2002) Carotid intima-media thickness, arterial stiffness and risk of cardiovascular disease: current evidence. J Hypertens 20:2317–2325
18. Furberg CD, Adams HP Jr, Applegate WB, Byington RP, Espeland MA, Hartwell T, Hunninghake DB, Lefkowitz DS, Probstfield J, Riley WA, et al (1994) Effect of lovastatin on early carotid atherosclerosis and cardiovascular events. Asymptomatic Carotid Artery Progression Study (ACAPS) Research Group. Circulation 90:1679–1687
19. Pitt B, Byington RP, Furberg CD, Hunninghake DB, Mancini GB, Miller ME, Riley W (2000) Effect of amlodipine on the progression of atherosclerosis and the occurrence of clinical events. PREVENT Investigators. Circulation 102:1503–1510
20. Salonen R, Nyyssonen K, Porkkala E, Rummukainen J, Belder R, Park JS, Salonen JT (1995) Kuopio Atherosclerosis Prevention Study (KAPS). A population-based primary preventive trial of the effect of LDL lowering on atherosclerotic progression in carotid and femoral arteries. Circulation 92:1758–1764
21. Smilde TJ, van Wissen S, Wollersheim H, Trip MD, Kastelein JJ, Stalenhoef AF (2001) Effect of aggressive versus conventional lipid lowering on atherosclerosis progression in familial hypercholesterolaemia (ASAP): a prospective, randomised, double-blind trial. Lancet 357:577–581

22. Terpstra WF, May JF, Smit AJ, Graeff PA, Meyboom-de Jong B, Crijns HJ (2004) Effects of amlodipine and lisinopril on intima-media thickness in previously untreated, elderly hypertensive patients (the ELVERA trial). J Hypertens 22:1309–1316
23. Zanchetti A, Bond MG, Hennig M, Neiss A, Mancia G, Dal Palu C, Hansson L, Magnani B, Rahn KH, Reid JL, Rodicio J, Safar M, Eckes L, Rizzini P (2002) Calcium antagonist lacidipine slows down progression of asymptomatic carotid atherosclerosis: principal results of the European Lacidipine Study on Atherosclerosis (ELSA), a randomized, double-blind, long-term trial. Circulation 106:2422–2427
24. Bots ML, Evans GW, Riley WA, Grobbee DE (2003) Carotid intima-media thickness measurements in intervention studies: design options, progression rates, and sample size considerations: a point of view. Stroke 34:2985–2994
25. Davis PH, Dawson JD, Riley WA, Lauer RM (2001) Carotid intimal-medial thickness is related to cardiovascular risk factors measured from childhood through middle age: the Muscatine Study. Circulation 104:2815–2819
26. Freedman DS, Dietz WH, Tang R, Mensah GA, Bond MG, Urbina EM, Srinivasan S, Berenson GS (2004) The relation of obesity throughout life to carotid intima-media thickness in adulthood: the Bogalusa Heart Study. Int J Obes Relat Metab Disord 28:159–166
27. Li S, Chen W, Srinivasan SR, Bond MG, Tang R, Urbina EM, Berenson GS (2003) Childhood cardiovascular risk factors and carotid vascular changes in adulthood: the Bogalusa Heart Study. JAMA 290:2271–2276
28. Oren A, Vos L, Uiterwaal C, Gorissen W, Grobbee D, Bots M (2003) Change in body mass index from adolescence to young adulthood and increased carotid intima-media thickness at 28 years of age: the atherosclerosis risk in young adults study. Int J Obes Relat Metab Disord 27:1383–1390
29. Raitakari OT, Juonala M, Kahonen M, Taittonen L, Laitinen T, Maki-Torkko N, Jarvisalo MJ, Uhari M, Jokinen E, Ronnemaa T, Akerblom HK, Viikari JS (2003) Cardiovascular risk factors in childhood and carotid artery intima-media thickness in adulthood: the cardiovascular risk in Young Finns Study. JAMA 290:2277–2283
30. Aggoun Y, Bonnet D, Sidi D, Girardet JP, Brucker E, Polak M, Safar ME, Levy BI (2000) Arterial mechanical changes in children with familial hypercholesterolemia. Arterioscler Thromb Vasc Biol 20:2070–2075
31. Jarvisalo MJ, Jartti L, Nanto-Salonen K, Irjala K, Ronnemaa T, Hartiala JJ, Celermajer DS, Raitakari OT (2001) Increased aortic intima-media thickness: a marker of preclinical atherosclerosis in high-risk children. Circulation 104:2943–2947
32. Koeijvoets KC, Rodenburg J, Hutten BA, Wiegman A, Kastelein JJ, Sijbrands EJ (2005) Low-density lipoprotein receptor genotype and response to pravastatin in children with familial hypercholesterolemia: substudy of an intima-media thickness trial. Circulation 112:3168–3173
33. Tonstad S, Joakimsen O, Stensland-Bugge E, Leren TP, Ose L, Russell D, Bonaa KH (1996) Risk factors related to carotid intima-media thickness and plaque in children with familial hypercholesterolemia and control subjects. Arterioscler Thromb Vasc Biol 16:984–991
34. Wiegman A, de Groot E, Hutten BA, Rodenburg J, Gort J, Bakker HD, Sijbrands EJ, Kastelein JJ (2004) Arterial intima-media thickness in children heterozygous for familial hypercholesterolaemia. Lancet 363:369–370
35. Lande MB, Carson NL, Roy J, Meagher CC (2006) Effects of childhood primary hypertension on carotid intima media thickness: a matched controlled study. Hypertension 48:40–44
36. Sorof JM, Alexandrov AV, Cardwell G, Portman RJ (2003) Carotid artery intimal-medial thickness and left ventricular hypertrophy in children with elevated blood pressure. Pediatrics 111:61–66
37. Meyer AA, Kundt G, Steiner M, Schuff-Werner P, Kienast W (2006) Impaired flow-mediated vasodilation, carotid artery intima-media thickening, and elevated endothelial plasma markers in obese children: the impact of cardiovascular risk factors. Pediatrics 117:1560–1567

38. Woo KS, Chook P, Yu CW, Sung RY, Qiao M, Leung SS, Lam CW, Metreweli C, Celermajer DS (2004) Overweight in children is associated with arterial endothelial dysfunction and intima-media thickening. Int J Obes Relat Metab Disord 28:852–857
39. Iannuzzi A, Licenziati MR, Acampora C, Salvatore V, Auriemma L, Romano ML, Panico S, Rubba P, Trevisan M (2004) Increased carotid intima-media thickness and stiffness in obese children. Diabetes Care 27:2506–2508
40. Singh TP, Groehn H, Kazmers A (2003) Vascular function and carotid intimal-medial thickness in children with insulin-dependent diabetes mellitus. J Am Coll Cardiol 41:661–665
41. Iannuzzi A, Licenziati MR, Acampora C, Renis M, Agrusta M, Romano L, Valerio G, Panico S, Trevisan M (2006) Carotid artery stiffness in obese children with the metabolic syndrome. Am J Cardiol 97:528–531
42. Sass C, Herbeth B, Chapet O, Siest G, Visvikis S, Zannad F (1998) Intima-media thickness and diameter of carotid and femoral arteries in children, adolescents and adults from the Stanislas cohort: effect of age, sex, anthropometry and blood pressure. J Hypertens 16:1593–1602
43. Ishizu T, Ishimitsu T, Yanagi H, Seo Y, Obara K, Moriyama N, Watanabe S, Yamaguchi I (2004) Effect of age on carotid arterial intima-media thickness in childhood. Heart Vessels 19:189–195
44. Espeland MA, Tang R, Terry JG, Davis DH, Mercuri M, Crouse JR III (1999) Associations of risk factors with segment-specific intimal-medial thickness of the extracranial carotid artery. Stroke 30:1047–1055
45. Jourdan C, Wuhl E, Litwin M, Fahr K, Trelewicz J, Jobs K, Schenk JP, Grenda R, Mehls O, Troger J, Schaefer F (2005) Normative values for intima-media thickness and distensibility of large arteries in healthy adolescents. J Hypertens 23:1707–1715
46. Urbina EM, Kimball TR, McCoy CE, Khoury PR, Daniels SR, Dolan LM (2009) Youth with obesity and obesity-related type 2 diabetes mellitus demonstrate abnormalities in carotid structure and function. Circulation 119:2913–2919
47. Koklu E, Kurtoglu S, Akcakus M, Yikilmaz A, Coskun A, Gunes T (2007) Intima-media thickness of the abdominal aorta of neonate with different gestational ages. J Clin Ultrasound 35:491–497
48. Skilton MR, Evans N, Griffiths KA, Harmer JA, Celermajer DS (2005) Aortic wall thickness in newborns with intrauterine growth restriction. Lancet 365:1484–1486
49. Gunes T, Koklu E, Yikilmaz A, Ozturk MA, Akcakus M, Kurtoglu S, Coskun A, Koklu S (2007) Influence of maternal smoking on neonatal aortic intima-media thickness, serum IGF-I and IGFBP-3 levels. Eur J Pediatr 166:1039–1044
50. Volanen I, Jarvisalo MJ, Vainionpaa R, Arffman M, Kallio K, Angle S, Ronnemaa T, Viikari J, Marniemi J, Raitakari OT, Simell O (2006) Increased aortic intima-media thickness in 11-year-old healthy children with persistent Chlamydia pneumoniae seropositivity. Arterioscler Thromb Vasc Biol 26:649–655
51. Volanen I, Kallio K, Saarinen M, Jarvisalo MJ, Vainionpaa R, Ronnemaa T, Viikari J, Marniemi J, Simell O, Raitakari OT (2008) Arterial intima-media thickness in 13-year-old adolescents and previous antichlamydial antimicrobial use: a retrospective follow-up study. Pediatrics 122:e675–e681
52. Dawson JD, Sonka M, Blecha MB, Lin W, Davis PH (2009) Risk factors associated with aortic and carotid intima-media thickness in adolescents and young adults: the Muscatine Offspring Study. J Am Coll Cardiol 53:2273–2279
53. Roman MJ, Saba PS, Pini R, Spitzer M, Pickering TG, Rosen S, Alderman MH, Devereux RB (1992) Parallel cardiac and vascular adaptation in hypertension. Circulation 86:1909–1918
54. Liang YL, Teede H, Kotsopoulos D, Shiel L, Cameron JD, Dart AM, McGrath BP (1998) Non-invasive measurements of arterial structure and function: repeatability, interrelationships and trial sample size. Clin Sci (Lond) 95:669–679
55. Wendelhag I, Wiklund O, Wikstrand J (1992) Arterial wall thickness in familial hypercholesterolemia. Ultrasound measurement of intima-media thickness in the common carotid artery. Arterioscler Thromb 12:70–77

56. Wendelhag I, Wiklund O, Wikstrand J (1995) Intima-media thickness after cholesterol lowering in familial hypercholesterolemia. A three-year ultrasound study of common carotid and femoral arteries. Atherosclerosis 117:225–236
57. Roman MJ, Pickering TG, Schwartz JE, Pini R, Devereux RB (1995) Association of carotid atherosclerosis and left ventricular hypertrophy. J Am Coll Cardiol 25:83–90
58. Riley W, Freedman D, Higgs N, Barnes R, Zinkgraf S, Berenson G (1992) Decreased arterial elasticity associated with cardiovascular disease risk factors in the young. The Bogalusa Heart Study. Arteriosclerosis 6:378–386
59. Cavallini M, Roman M, Blank S, Pini R, Pickering T, Devereux R (1996) Association of the auscultatory gap with vascular disease in hypertensive patients. Ann Intern Med 124:877–883
60. Salomaa V, Riley W, Kark JD, Nardo C, Folsom AR (1995) Non-insulin-dependent diabetes mellitus and fasting glucose and insulin concentrations are associated with arterial stiffness indexes. The ARIC Study. Atherosclerosis Risk in Communities Study. Circulation 91:1432–1443
61. Urbina EM, Srinivasan, SR, Kieltyka RL, Tang R, Bond MG, Chen W, Berenson GS (2004) Correlates of carotid artery stiffness in young adults: the Bogalusa Heart Study. Atherosclerosis 176:157–164
62. Urbina EM, Williams RV, Alpert BS, Collins RT, Daniels SR, Hayman L, Jacobson M, Mahoney L, Mietus-Snyder M, Rocchini A, Steinberger J, McCrindle B (2009) on behalf of the American Heart Association Atherosclerosis H, Obesity in Youth Committee of the Council on Cardiovascular Disease in the Young. Noninvasive assessment of subclinical atherosclerosis in children and adolescents. Recommendations for standard assessment for clinical research. A Scientific Statement from the American Heart Association. Hypertension. HYPERTENSIONAHA.109.192639
63. Woo KS, Chook P, Yu CW, Sung RY, Qiao M, Leung SS, Lam CW, Metreweli C, Celermajer DS (2004) Effects of diet and exercise on obesity-related vascular dysfunction in children. Circulation 109:1981–1986
64. Wiegman A, Hutten BA, de Groot E, Rodenburg J, Bakker HD, Buller HR, Sijbrands EJ, Kastelein JJ (2004) Efficacy and safety of statin therapy in children with familial hypercholesterolemia: a randomized controlled trial. J Am Med Assoc 292:331–337
65. Crouse JR III, Raichlen JS, Riley WA, Evans GW, Palmer MK, O'Leary DH, Grobbee DE, Bots ML (2007) Effect of rosuvastatin on progression of carotid intima-media thickness in low-risk individuals with subclinical atherosclerosis: the METEOR Trial. J Am Med Assoc 297:1344–1353
66. Anderson HC (1983) Calcific diseases. Arch Path Lab Med 107:341–348
67. Stary HC (1989) Evolution and progression of atherosclerotic lesions in coronary arteries of children and young adults. Arteriosclerosis 9(Suppl I):I-19–I-32
68. Epstein FH, Simpson R, Boas EP (1957) The epidemiology of atherosclerosis among a random sample of clothing workers of different ethnic origins in New York City. II. Associations between serum lipid levels, blood pressure, overweight, and some other variables. J Chron Dis 5:329–341
69. Margolis JR, Chen JTT, Kong Y, Peter RH, Behar VS, Kisslo JA (1980) The diagnostic and prognostic significance of coronary artery calcification. Radiology 137:609–616
70. Wang L, Jerosch-Herold M, Jacobs DR Jr, Shahar E, Detrano R, Folsom AR (2006) Coronary artery calcification and myocardial perfusion in asymptomatic adults: the MESA (Multi-Ethnic Study of Atherosclerosis). J Am Coll Cardiol 48:1018–1026
71. Mahoney LT, Burns TL, Stanford W, Thompson BH, Witt JD, Rost CA, Lauer RM (1996) Coronary risk factors measured in childhood and young adult life are associated with coronary artery calcification in young adults: the Muscatine Study. J Am Coll Cardiol 27:277–284
72. Janowitz WR, Agatston AS, Kaplan G, Viamonte M Jr (1993) Differences in prevalence and extent of coronary artery calcium detected by ultrafast computed tomography in asymptomatic men and women. Am J Cardiol 72:247–254

73. Sakuma H, Takeda K, Hirano T, Okuda Y, Matsumura K, Yamaguchi N, Nakana T, Takano M (1988) Plain chest radiograph with computed tomography; improved sensitivity for the detection of coronary artery calcification. Am J Roentgen 151:27–30
74. Bormann J, Stanford W, Stenberg R, Winniford M, Galvin J, Berbaum K, Talman C (1992) Ultrafast CT detection of coronary artery stenosis. Am J Cardiac Imaging 6:191–196
75. Hamby RI, Tabrah F, Wisoff BG, Hartstein ML (1974) Coronary artery calcification: clinical implications and angiographic correlates. Am Heart J 87:565–570
76. Wong ND, Kouwabunpat D, Vo A, Detrano RC, Eisenberg H, Goel M, Tobis JM (1994) Coronary calcium and atherosclerosis by ultrafast computed tomography in asymptomatic men and women: relation to age and risk factors. Am Heart J 127:422–430
77. Wong ND, Vo A, Abrahamson D, Tobis JM, Eisenberg H, Detrano RC (1994) Detection of coronary artery calcium by ultrafast computed tomography and its relation to clinical evidence of coronary artery disease. Am J Cardiol 73:223–227
78. Goel M, Wong ND, Eisenberg H, Hagar J, Kelly D, Tobis JM (1992) Risk factor correlates of coronary calcium as evaluated by ultrafast computed tomography. Am J Cardiol 70:977–980
79. Loecker TH, Schwartz RS, Cotta CW, Hickman JR Jr (1992) Fluoroscopic coronary artery calcification and associated coronary disease in asymptomatic young men. J Am Coll Cardiol 19:1167–1172
80. Michos ED, Nasir K, Braunstein JB, Rumberger JA, Budoff MJ, Post WS, Blumenthal RS (2006) Framingham risk equation underestimates subclinical atherosclerosis risk in asymptomatic women. Atherosclerosis 184:201–206
81. Dabelea D, Kinney G, Snell-Bergeon JK, Hokanson JE, Eckel RH, Ehrlich J, Garg S, Hamman RF, Rewers M (2003) Effect of Type 1 diabetes on the gender difference in coronary artery calcification: a role for Insulin resistance?: the Coronary Artery Calcification in Type 1 Diabetes (CACTI) Study. Diabetes 52:2833–2839
82. Gidding SS, Bookstein LC, Chomka EV (1998) Usefulness of electron beam tomography in adolescents and young adults with heterozygous familial hypercholesterolemia. Circulation 98:2580–2583
83. Kaichi S, Tsuda E, Fujita H, Kurosaki K, Tanaka R, Naito H, Echigo S (2008) Acute coronary artery dilation due to Kawasaki disease and subsequent late calcification as detected by electron beam computed tomography. Pediatr Cardiol 29:568–573
84. Frey EE, Matherne CP, Mahoney LT, Sato Y, Stanford W, Smith WL (1988) Coronary artery aneurysms due to Kawasaki disease: diagnosis with ultrafast CT. Radiology 167:725–726
85. Dadlani GH, Gingell RL, Orie JD, Roland JM, Najdzionek J, Lipsitz SR, Pieroni DR, Lipshultz SE (2005) Coronary artery calcifications in the long-term follow-up of Kawasaki disease. Am Heart J 150:1016
86. Goodman WG, Goldin J, Kuizon BD, Yoon C, Gales B, Sider D, Wang Y, Chung J, Emerick A, Greaser L, Elashoff RM, Salusky IB (2000) Coronary-artery calcification in young adults with end-stage renal disease who are undergoing dialysis. N Engl J Med 342:1478–1483
87. Sheth RD, Perez MD, goldstein SL (2003) Cardiovascular calcifications in pediatric patients receiving maintenance dialysis. Pediatr Nephrol 18:810–813
88. Ishitani MB, Milliner DS, Kim DY, Bohorquez HE, Heimbach JK, Sheedy PF II, Morgenstern BZ, Gloor JM, Murphy JG, McBane RD, Bielak LF, Peyser PA, Stegall MD (2005) Early subclinical coronary artery calcification in young adults who were pediatric kidney transplant recipients. Am J Transplant 5:1689–1693
89. Bae KT, Hong C, Takahashi N, Gutierrez F, Sharkey AM, Hirsch R, Canter CE, Bae KT, Hong C, Takahashi N, Gutierrez F, Sharkey AM, Hirsch R, Canter CE (2004) Multi-detector row computed tomographic angiography in pediatric heart transplant recipients: initial observations. Transplantation 77:599–602
90. Starkman HS, Cable G, Hala V, Hecht H, Donnelly CM, Starkman HS, Cable G, Hala V, Hecht H, Donnelly CM (2003) Delineation of prevalence and risk factors for early coronary artery disease by electron beam computed tomography in young adults with type 1 diabetes. Diabetes Care 26:433–436

91. O'Rourke MF, Staessen JA, Vlachopoulos C, Duprez D, Plante GE (2002) Clinical applications of arterial stiffness; definitions and reference values. Am J Hypertens 15:426–444
92. Laurent S, Hayoz D, Trazzi S, Boutouyrie P, Waeber B, Omboni S, Brunner HR, Mancia G, Safar M (1993) Isobaric compliance of the radial artery is increased in patients with essential hypertension. J Hypertens 11:89–98
93. Mangoni AA, Mircoli L, Giannattasio C, Ferrari AU, Mancia G (1996) Heart rate-dependence of arterial distensibility in vivo. J Hypertens 14:897–901
94. Safar ME, London BM (1994) The arterial system in human hypertension. Blackwell Scientific, London
95. Bank AJ, Kaiser DR, Rajala S, Cheng A (1999) In vivo human brachial artery elastic mechanics: effects of smooth muscle relaxation. Circulation 100:41–47
96. Bank AJ, Wang H, Holte JE, Mullen K, Shammas R, Kubo SH (1996) Contribution of collagen, elastin, and smooth muscle to in vivo human brachial artery wall stress and elastic modulus. Circulation 94:3263–3270
97. Lopez-Beltran EA, Blackshear PL, Finkelstein SM, Cohn JN (1998) Non-invasive studies of peripheral vascular compliance using a non-occluding photoplethysmographic method. Med Biol Eng Comput 36:748–753
98. Giannattasio C, Mangoni AA, Failla M, Stella ML, Carugo S, Bombelli M, Sega R, Mancia G (1997) Combined effects of hypertension and hypercholesterolemia on radial artery function. Hypertension 29:583–586
99. Hopkins KD, Lehmann ED, Jones RL, Turay RC, Gosling RG (1996) A family history of NIDDM is associated with decreased aortic distensibility in normal healthy young adult subjects. Diabetes Care 19:501–503
100. Urbina EM, Brinton TJ, Elkasabany A, Berenson GS (2002) Brachial artery distensibility and relation to cardiovascular risk factors in healthy young adults (The Bogalusa Heart Study). Am J Cardiol 89:946–951
101. Urbina EM, Kieltkya L, Tsai J, Srinivasan SR, Berenson GS (2005) Impact of multiple cardiovascular risk factors on brachial artery distensibility in young adults: the Bogalusa Heart Study. Am J Hypertens 18:767–771
102. Liao D, Arnett DK, Tyroler HA, Riley WA, Chambless LE, Szklo M, Heiss G (1999) Arterial stiffness and the development of hypertension. The ARIC study. Hypertension 34:201–206
103. Cruickshank K, Riste L, Anderson SG, Wright JS, Dunn G, Gosling RG (2002) Aortic pulse-wave velocity and its relationship to mortality in diabetes and glucose intolerance: an integrated index of vascular function? Circulation 106:2085–2090
104. Laurent S, Boutouyrie P, Asmar R, Gautier I, Laloux B, Guize L, Ducimetiere P, Benetos A (2001) Aortic stiffness is an independent predictor of all-cause and cardiovascular mortality in hypertensive patients. Hypertension 37:1236–1241
105. Blacher J, Asmar R, Diane S, London GM, Safar ME (1999) Aortic pulse wave velocity as a marker of cardiovascular risk in hypertensive patients. Hypertension 33:1111–1117
106. Syeda B, Gottsauner-Wolf M, Denk S, Pichler P, Khorsand A, Glogar D (2003) Arterial compliance: a diagnostic marker for atherosclerotic plaque burden? Am J Hypertens 16:356–362
107. Bhuiyan AR, Li S, Li H, Chen W, Srinivasan SR, Berenson GS (2005) Distribution and correlates of arterial compliance measures in asymptomatic young adults: the Bogalusa Heart Study. Am J Hypertens 18:684–691
108. Li H, Srinivasan SR, Chen W, Xu JH, Li S, Berenson GS (2005) Vascular abnormalities in asymptomatic, healthy young adult smokers without other major cardiovascular risk factors: the Bogalusa Heart Study. Am J Hypertens 18:319–324
109. Li H, Srinivasan SR, Berenson GS (2006) Comparison of the measures of pulsatile arterial function between asymptomatic younger adult smokers and former smokers: the Bogalusa Heart Study. Am J Hypertens 19:897–901

110. Arnett DK, Glasser SP, McVeigh G, Prineas R, Finklestein S, Donahue R, Cohn JN, Sinaiko A (2001) Blood pressure and arterial compliance in young adults: the Minnesota Children's Blood Pressure Study. Am J Hypertens 14:200–205
111. Covic A, Haydar AA, Bhamra-Ariza P, Gusbeth-Tatomir P, Goldsmith DJ (2005) Aortic pulse wave velocity and arterial wave reflections predict the extent and severity of coronary artery disease in chronic kidney disease patients. J Nephrol 18:388–396
112. Jelic S, Bartels MN, Mateika JH, Ngai P, DeMeersman RE, Basner RC (2002) Arterial stiffness increases during obstructive sleep apneas. Sleep 25:850–855
113. London GM, Blacher J, Pannier B, Guerin AP, Marchais SJ, Safar ME (2001) Arterial wave reflections and survival in end-stage renal failure. Hypertension 38:434–438
114. Wilkinson IB, Prasad K, Hall IR, Thomas A, MacCallum H, Webb DJ, Frenneaux MP, Cockcroft JR (2002) Increased central pulse pressure and augmentation index in subjects with hypercholesterolemia. J Am Coll Cardiol 39:1005–1011
115. Wilkinson IB, MacCallum H, Rooijmans DF, Murray GD, Cockcroft JR, McKnight JA, Webb DJ (2000) Increased augmentation index and systolic stress in type 1 diabetes mellitus. QJM 93:441–448
116. Tomiyama H, Arai T, Koji Y, Yambe M, Motobe K, Zaydun G, Yamamoto Y, Hori S, Yamashina A (2004) The age-related increase in arterial stiffness is augmented in phases according to the severity of hypertension. Hypertens Res-Clin Exp 27:465–470
117. Schnabel R, Larson MG, Dupuis J, Lunetta KL, Lipinska I, Meigs JB, Yin X, Rong J, Vita JA, Newton-Cheh C, Levy D, Keaney JF Jr, Vasan RS, Mitchell GF, Benjamin EJ (2008) Relations of inflammatory biomarkers and common genetic variants with arterial stiffness and wave reflection. Hypertension 51:1651–1657
118. Kohara K, Tabara Y, Oshiumi A, Miyawaki Y, Kobayashi T, Miki T (2005) Radial augmentation index: a useful and easily obtainable parameter for vascular aging. Am J Hypertens 18:11S–14S
119. Matsui Y, Ishikawa J, Eguchi K, Hoshide S, Miyashita H, Shimada K, Kario K, Matsui Y, Ishikawa J, Eguchi K, Hoshide S, Miyashita H, Shimada K, Kario K (2008) The influence of wave reflection on left ventricular hypertrophy in hypertensive patients is modified by age and gender. Hypertens Res-Clin Exp 31:649–656
120. Olafiranye O, Qureshi G, Salciccioli L, Weber M, Lazar J (2008) Association of beta-blockers use with increased aortic wave reflection. J Am Soc Hypertens 2:64–69
121. de Luca N, Mallion JM, O'Rourke MF, O'Brien E, Rahn KH, Trimarco B, Romero R, De Leeuw PW, Hitzenberger G, Battegay E, Duprez D, Sever P, Safar ME (2004) Regression of left ventricular mass in hypertensive patients treated with perindopril/indapamide as a first-line combination: the REASON echocardiography study. Am J Hypertens 17:660–667
122. Papamichael CM, Karatzi KN, Papaioannou TG, Karatzis EN, Katsichti P, Sideris V, Zakopoulos N, Zampelas A, Lekakis JP (2008) Acute combined effects of olive oil and wine on pressure wave reflections: another beneficial influence of the Mediterranean diet antioxidants? J Hypertens 26:223–229
123. Lehmann ED, Gosling RG, Sonksen PH (1992) Arterial wall compliance in diabetes. Diabet Med 9:114–119
124. Nichols WW, O'Rourke MF (1990) McDonald's blood flow in arteries: theoretical, experimental and clinical principles, 3rd edn. Edward Arnold, London
125. London GM, Marchais SJ, Safar ME, Genest AF, Guerin AP, Metivier F, Chedid K, London AM (1990) Aortic and large artery compliance in end-stage renal failure. Kidney Int 37:137–142
126. Lehmann ED, Watts GF, Gosling RG (1992) Aortic distensibility and hypercholesterolaemia. Lancet 340:1171–1172
127. Avolio AP, Chen SG, Wang RP, Zhang CL, Li MF, O'Rourke MF (1983) Effects of aging on changing arterial compliance and left ventricular load in a northern Chinese urban community. Circulation 68:50–58

128. Vaitkevicius PV, Fleg JL, Engel JH, O'Connor FC, Wright JG, Lakatta LE, Yin FC, Lakatta EG (1993) Effects of age and aerobic capacity on arterial stiffness in healthy adults. Circulation 88:1456–1462
129. Li S, Chen W, Srinivasan SR, Berenson GS (2005) Influence of metabolic syndrome on arterial stiffness and its age-related change in young adults: the Bogalusa Heart Study. Atherosclerosis 180:349–354
130. Li S, Chen W, Srinivasan SR, Berenson GS (2004) Childhood blood pressure as a predictor of arterial stiffness in young adults: the bogalusa heart study. Hypertension 43:541–546
131. Whincup PH, Gilg JA, Donald AE, Katterhorn M, Oliver C, Cook DG, Deanfield JE (2005) Arterial distensibility in adolescents: the influence of adiposity, the metabolic syndrome, and classic risk factors. Circulation 112:1789–1797
132. Leeson CP, Whincup PH, Cook DG, Mullen MJ, Donald AE, Seymour CA, Deanfield JE (2000) Cholesterol and arterial distensibility in the first decade of life: a population-based study. Circulation 101:1533–1538
133. Singhal A, Farooqi IS, Cole TJ, O'Rahilly S, Fewtrell M, Kattenhorn M, Lucas A, Deanfield J (2002) Influence of leptin on arterial distensibility: a novel link between obesity and cardiovascular disease? Circulation 106:1919–1924
134. Urbina EM, Bean JA, Daniels SR, D'Alessio D, Dolan, LM (2007) Overweight and hyperinsulinemia provide individual contributions to compromises in brachial artery distensibility in healthy adolescents and young adults: brachial distensibility in children. J Am Soc Hypertens 1:200–207
135. Urbina EM, Khoury P, Martin LJ, D'Alessio D, Dolan LM (2009) Gender differences in the relationships among obesity, adiponectin and brachial artery distensibility in adolescents and young adults. Int J Obes 33:1118–1125
136. Haller MJ, Samyn M, Nichols WW, Brusko T, Wasserfall C, Schwartz RF, Atkinson M, Shuster JJ, Pierce GL, Silverstein JH (2004) Radial artery tonometry demonstrates arterial stiffness in children with type 1 diabetes. Diabetes Care 27:2911–2917
137. Galler A, Heitmann A, Siekmeyer W, Gelbrich G, Kapellen T, Kratzsch J, Kiess W (2009) Increased arterial stiffness in children and adolescents with type 1 diabetes: no association between arterial stiffness and serum levels of adiponectin. Pediatr Diabetes 11:38–46
138. Brooks BA, Molyneaux LM, Yue DK (2001) Augmentation of central arterial pressure in Type 2 diabetes. Diabet Med 18:374–380
139. Covic A, Mardare N, Gusbeth-Tatomir P, Brumaru O, Gavrilovici C, Munteanu M, Prisada O, Goldsmith DJ (2006) Increased arterial stiffness in children on haemodialysis. Nephrol Dial Transplant 21:729–735
140. Mahmud A, Feely J (2003) Spurious systolic hypertension of youth: fit young men with elastic arteries. Am J Hypertens 16:229–232
141. Kyvelou S-MG, Vyssoulis GP, Karpanou EA, Adamopoulos DN, Deligeorgis AD, Cokkinos DV, Stefanadis CI (2008) Arterial stiffness in offspring of hypertensive parents: a pilot study. Int J Cardiol 129:438–440
142. Kucerova J, Filipovsky J, Staessen JA, Cwynar M, Wojciechowska W, Stolarz K, Kuznetsova T, Gasowski J, Dolejsova M, Grodzicki T, Kawecka-Jaszcz K, Fagard R (2006) Arterial characteristics in normotensive offspring of parents with or without a history of hypertension. Am J Hypertens 19:264–269
143. Stella A, Gessaroli M, Cifiello BI, Salardi S, Reggiani A, Cacciari E, D'Addato M (1984) Elastic modulus in young diabetic patients (ultrasound measurements of pulse wave velocity). Angiology 35:729–734
144. Woolam GL, Schnur PL, Vallbona C, Hoff HE (1962) The pulse wave velocity as an early indicator of atherosclerosis in diabetic subjects. Circulation 25:533–539
145. Tedesco MA, Di Salvo G, Ratti G, Natale F, Calabrese E, Grassia C, Iacono A, Lama G (2001) Arterial distensibility and ambulatory blood pressure monitoring in young patients with neurofibromatosis type 1. Am J Hypertens 14:559–566

146. Tedesco MA, Ratti G, Di Salvo G, Martiniello AR, Limongelli G, Grieco M, Calabrese E, Grassia C, Iacono A, Lama G (2000) Noninvasive evaluation of arterial abnormalities in young patients with neurofibromatosis type 1. Angiology 51:733–741
147. Kwok KL, Ng DK, Cheung YF (2003) BP and arterial distensibility in children with primary snoring. Chest 123:1561–1566
148. Cheung YF, Yung TC, Tam SC, Ho MH, Chau AK (2004) Novel and traditional cardiovascular risk factors in children after Kawasaki disease: implications for premature atherosclerosis. J Am Coll Cardiol 43:120–124
149. Cheung YF, Brogan PA, Pilla CB, Dillon MJ, Redington AN (2002) Arterial distensibility in children and teenagers: normal evolution and the effect of childhood vasculitis. Arch Dis Child 87:348–351
150. de Divitiis M, Pilla C, Kattenhorn M, Donald A, Zadinello M, Wallace S, Redington A, Deanfield J (2003) Ambulatory blood pressure, left ventricular mass, and conduit artery function late after successful repair of coarctation of the aorta. J Am Coll Cardiol 41:2259–2265
151. Roegel JC, Heinrich E, De Jong W, Stephan D, Charpentier A, Eisenmann B, Imbs JL (1998) Vascular and neuroendocrine components in altered blood pressure regulation after surgical repair of coarctation of the aorta. J Hum Hypertens 12:517–525
152. Ou P, Celermajer DS, Raisky O, Jolivet O, Buyens F, Herment A, Sidi D, Bonnet D, Mousseaux E (2008) Angular (Gothic) aortic arch leads to enhanced systolic wave reflection, central aortic stiffness, and increased left ventricular mass late after aortic coarctation repair: evaluation with magnetic resonance flow mapping. J Thorac Cardiovasc Surg 135:62–68
153. Im JA, Lee JW, Shim JY, Lee HR, Lee DC (2007) Association between brachial-ankle pulse wave velocity and cardiovascular risk factors in healthy adolescents. J Pediatr 150:247–251
154. Zhu H, Yan W, Ge D, Treiber FA, Harshfield GA, Kapuku G, Snieder H, Dong Y (2007) Cardiovascular characteristics in American youth with prehypertension. Am J Hypertens 20:1051–1057
155. Ge D, Young TW, Wang X, Kapuku GK, Treiber FA, Snieder H (2007) Heritability of arterial stiffness in black and white American youth and young adults. Am J Hypertens 20:1065–1072
156. Urbina EM, Kimball TR et al (2010) Increased arterial stiffness is found in adolescents with obesity or obesity-related type 2 diabetes mellitus. J Hypertens 28(8):1692–1698
157. Avolio AP, Deng FQ, Li WQ, Luo YF, Huang ZD, Xing LF, O'Rourke MF (1985) Effects of aging on arterial distensibility in populations with high and low prevalence of hypertension: comparison between urban and rural communities in China. Circulation 71:202–210
158. Okada M, Yakata M (1989) Normal ranges of pulse wave velocity and the changes with age. Rinsho Byori 37:215–218
159. McEniery CM, Yasmin, Hall IR, Qasem A, Wilkinson IB, Cockcroft JR (2005) Normal vascular aging: differential effects on wave reflection and aortic pulse wave velocity: the Anglo-Cardiff Collaborative Trial (ACCT). J Am Coll Cardiol 46:1753–1760
160. Collins RT, Somes GW, Alpert BS (2008) Differences in arterial compliance among normotensive adolescent groups: Collins arterial compliance in adolescents. Pediatr Cardiol 29:929–934
161. Niboshi A, Hamaoka K, Sakata K, Inoue F (2006) Characteristics of brachial-ankle pulse wave velocity in Japanese children. Eur J Pediatr 165:625–629
162. Urbina EM, Dolan LM, McCoy CE, Khoury PR, Daniels SR, Kimball TR (2011) Relationship between elevated arterial stiffness and increased left ventricular mass in adolescents and young adults. J Pediatr 158:715–721
163. Deanfield J, Donald A, Ferri C, Giannattasio C, Halcox J, Halligan S, Lerman A, Mancia G, Oliver JJ, Pessina AC, Rizzoni D, Rossi GP, Salvetti A, Schiffrin EL, Taddei S, Webb DJ (2005) Endothelial function and dysfunction. Part I: methodological issues for assessment in the different vascular beds: a statement by the working group on endothelin and endothelial factors of the European society of hypertension. J Hypertens 23:7–17

164. Chan SY, Mancini GB, Kuramoto L, Schulzer M, Frohlich J, Ignaszewski A (2003) The prognostic importance of endothelial dysfunction and carotid atheroma burden in patients with coronary artery disease. J Am Coll Cardiol 42:1037–1043
165. Li XP, Zhao SP, Zhang XY, Liu L, Gao M, Zhou QC (2000) Protective effect of high density lipoprotein on endothelium-dependent vasodilatation. Int J Cardiol 73:231–236
166. Felmeden DC, Spencer CG, Blann AD, Beevers DG, Lip GY (2003) Low-density lipoprotein subfractions and cardiovascular risk in hypertension: relationship to endothelial dysfunction and effects of treatment. Hypertension 41:528–533
167. Balletshofer BM, Rittig K, Stock J, Lehn-Stefan A, Overkamp D, Dietz K, Haring HU (2003) Insulin resistant young subjects at risk of accelerated atherosclerosis exhibit a marked reduction in peripheral endothelial function early in life but not differences in intima-media thickness. Atherosclerosis 171(2):303–309
168. Lee SJ, Lee DW, Kim KS, Lee IK (2001) Effect of estrogen on endothelial dysfunction in postmenopausal women with diabetes. Diabetes Res Clin Pract 54:S81–S92
169. Celermajer DS, Adams MR, Clarkson P, Robinson J, McCredie R, Donald A, Deanfield JE (1996) Passive smoking and impaired endothelium-dependent arterial dilatation in healthy young adults. N Engl J Med 334(3):150–154
170. Franzoni F, Ghiadoni L, Galetta F, Plantinga Y, Lubrano V, Huang Y, Salvetti G, Regoli F, Taddei S, Santoro G, Salvetti A (2005) Physical activity, plasma antioxidant capacity, and endothelium-dependent vasodilation in young and older men. Am J Hypertens 18(4 Pt 1):510–516
171. Amar S, Gokce N, Morgan S, Loukideli M, Van Dyke TE, Vita JA (2003) Periodontal disease is associated with brachial artery endothelial dysfunction and systemic inflammation. Arterioscler Thromb Vasc Biol 23:1245–1249
172. Ohike Y, Kozaki K, Iijima K, Eto M, Kojima T, Ohga E, Santa T, Imai K, Hashimoto M, Yoshizumi M, Ouchi Y (2005) Amelioration of vascular endothelial dysfunction in obstructive sleep apnea syndrome by nasal continuous positive airway pressure—possible involvement of nitric oxide and asymmetric NG, NG-dimethylarginine. Circ J 69:221–226
173. Ip MS, Tse HF, Lam B, Tsang KW, Lam WK (2004) Endothelial function in obstructive sleep apnea and response to treatment. Am J Respir Crit Care Med 169:348–353
174. Fuchsjager-Mayrl G, Pleiner J, Wiesinger GF, Sieder AE, Quittan M, Nuhr MJ, Francesconi C, Seit HP, Francesconi M, Schmetterer L, Wolzt M (2002) Exercise training improves vascular endothelial function in patients with type 1 diabetes. Diabetes Care 25:1795–1801
175. Hamdy O, Ledbury S, Mullooly C, Jarema C, Porter S, Ovalle K, Moussa A, Caselli A, Caballero AE, Economides PA, Veves A, Horton ES (2003) Lifestyle modification improves endothelial function in obese subjects with the insulin resistance syndrome. Diabetes Care 26:2119–2125
176. Ryan M, McInerney D, Owens D, Collins P, Johnson A, Tomkin GH (2000) Diabetes and the Mediterranean diet: a beneficial effect of oleic acid on insulin sensitivity, adipocyte glucose transport and endothelium-dependent vasoreactivity. QJM 93:85–91
177. Sondergaard E, Moller JE, Egstrup K (2003) Effect of dietary intervention and lipid-lowering treatment on brachial vasoreactivity in patients with ischemic heart disease and hypercholesterolemia. Am Heart J 145:E19
178. Williams MJ, Sutherland WH, McCormick MP, de Jong SA, McDonald JR, Walker RJ (2001) Vitamin C improves endothelial dysfunction in renal allograft recipients. Nephrol Dial Transplant 16:1251–1255
179. Agewall S, Wright S, Doughty RN, Whalley GA, Duxbury M, Sharpe N (2000) Does a glass of red wine improve endothelial function? Eur Heart J 21:74–78
180. Chou EJ, Keevil JG, Aeschlimann S, Wiebe DA, Folts JD, Stein JH (2001) Effect of ingestion of purple grape juice on endothelial function in patients with coronary heart disease. Am J Cardiol 88:553–555
181. Vlachopoulos C, Aznaouridis K, Alexopoulos N, Economou E, Andreadou I, Stefanadis C (2005) Effect of dark chocolate on arterial function in healthy individuals. Am J Hypertens 18:785–791

182. Woo KS, Chook P, Lolin YI, Sanderson JE, Metreweli C, Celermajer DS (1999) Folic acid improves arterial endothelial function in adults with hyperhomocystinemia. J Am Coll Cardiol 34:2002–2006
183. Alonso R, Mata P, De Andres R, Villacastin BP, Martinez-Gonzalez J, Badimon L (2001) Sustained long-term improvement of arterial endothelial function in heterozygous familial hypercholesterolemia patients treated with simvastatin. Atherosclerosis 157:423–429
184. de Jongh S, Lilien MR, op't Roodt J, Stroes ES, Bakker HD, Kastelein JJ (2002) Early statin therapy restores endothelial function in children with familial hypercholesterolemia. J Am Coll Cardiol 40:2117–2121
185. Joyce M, Moore K, Thompson C, Fitzgerald P, Fennessy F, Kelly CJ, Bouchier-Hayes DJ (2004) Hydroxy-methylglutaryl-coenzyme A reductase inhibition improves endothelial dysfunction in type-1 diabetes. Eur J Vasc Endovasc Surg 27:432–437
186. Beckman JA, Liao JK, Hurley S, Garrett LA, Chui D, Mitra D, Creager MA (2004) Atorvastatin restores endothelial function in normocholesterolemic smokers independent of changes in low-density lipoprotein. Circ Res 95:217–223
187. Ghiadoni L, Huang Y, Magagna A, Buralli S, Taddei S, Salvetti A (2001) Effect of acute blood pressure reduction on endothelial function in the brachial artery of patients with essential hypertension. J Hypertens 19:547–551
188. Anderson TJ, Elstein E, Haber H, Charbonneau F (2000) Comparative study of ACE-inhibition, angiotensin II antagonism, and calcium channel blockade on flow-mediated vasodilation in patients with coronary disease (BANFF study). J Am Coll Cardiol 35:60–66
189. Jensen-Urstad K, Johansson J (2001) Gender difference in age-related changes in vascular function. J Intern Med 250:29–36
190. Atkov OY, Balahonova TV, Pogorelova OA (1998) Non-invasive ultrasound detection of endothelial dysfunction. Eur J Ultrasound 7:37–45
191. Tounian P, Aggoun Y, Dubem B, Varille V, Guy-Grand B, Sidi D, Giradet J-P, Bonnet D (2001) Presence of increased stiffness of the common carotid artery and endothelial dysfunction in severely obese children: a prospective study. Lancet 358:1400–1404
192. Bonnet D, Aggoun Y, Szezepanski I, Bellal N, Blanche S (2004) Arterial stiffness and endothelial dysfunction in HIV-infected children. AIDS 18:1037–1041
193. Kari JA, Donald AE, Vallance DT, Bruckdorfer KR, Leone A, Mullen MJ, Bunce T, Dorado B, Deanfield JE, Rees L (1997) Physiology and biochemistry of endothelial function in children with chronic renal failure. Kidney Int 52:468–472
194. Deng YB, Li TL, Xiang HJ, Chang Q, Li CL (2003) Impaired endothelial function in the brachial artery after Kawasaki disease and the effects of intravenous administration of vitamin C. Pediatr Infect Dis J 22:34–39
195. Jarvisalo MJ, Lehtimaki T, Raitakari OT (2004) Determinants of arterial nitrate-mediated dilatation in children: role of oxidized low-density lipoprotein, endothelial function, and carotid intima-media thickness. Circulation 109:2885–2889
196. Charakida M, Donald AE, Green H, Storry C, Clapson M, Caslake M, Dunn DT, Halcox JP, Gibb DM, Klein NJ, Deanfield JE (2005) Early structural and functional changes of the vasculature in HIV-infected children: impact of disease and antiretroviral therapy. Circulation 112:103–109
197. Celermajer DS, Sorensen KE, Gooch VM, Spiegelhalter DJ, Miller OI, Sullivan ID, Lloyd JK, Deanfield JE (1992). Non-invasive detection of endothelial dysfunction in children and adults at risk of atherosclerosis. Lancet 340(8828):1111–1115
198. Jarvisalo MJ, Raitakari M, Toikka JO, Putto-Laurila A, Rontu R, Laine S, Lehtimaki T, Ronnemaa T, Viikari J, Raitakari OT (2004) Endothelial dysfunction and increased arterial intima-media thickness in children with type 1 diabetes. Circulation 109:1750–1755
199. Donaghue KC, Robinson J, McCredie R, Fung A, Silink M, Celermajer DS (1998) Macroangiopathy—does it play a role in young people? Horm Res 1:38–40
200. Lilien MR, Stroes ES, Op't Roodt J, de Jongh S, Schroder CH, Koomans HA (2003) Vascular function in children after renal transplantation. Am J Kidney Dis 41:684–691

201. Jarvisalo MJ, Harmoinen A, Hakanen M, Paakkunainen U, Viikari J, Hartiala J, Lehtimaki T, Simell O, Raitakari OT (2002) Elevated serum C-reactive protein levels and early arterial changes in healthy children. Arterioscler Thromb Vasc Biol 22:1323–1328
202. Aburawi E, Liuba P, Pesonen E, Yla-Herttuala S, Sjoblad S (2004) Acute respiratory viral infections aggravate arterial endothelial dysfunction in children with type 1 diabetes. Diabetes Care 27:2733–2735
203. Donaghue KC, Robinson J, McCredie R, Fung A, Silink M, Celermajer DS (1997) Large vessel dysfunction in diabetic adolescents and its relationship to small vessel complications. J Pediatr Endocrinol Metab 10:593–598
204. de Jongh S, Lilien MR, Bakker HD, Hutten BA, Kastelein JJ, Stroes ES (2002) Family history of cardiovascular events and endothelial dysfunction in children with familial hypercholesterolemia. Atherosclerosis 163:193–197
205. Mietus-Snyder M, Malloy MJ (1998) Endothelial dysfunction occurs in children with two genetic hyperlipidemias: improvement with antioxidant vitamin therapy. J Pediatr 133:35–40
206. Sorensen KE, Celermajer DS, Georgakopoulos D, Hatcher G, Betteridge DJ, Deanfield JE (1994) Impairment of endothelium-dependent dilation is an early event in children with familial hypercholesterolemia and is related to the lipoprotein(a) level. J Clin Invest 93:50–55
207. Aggoun Y, Farpour-Lambert NJ, Marchand LM, Golay E, Maggio AB, Beghetti M (2008) Impaired endothelial and smooth muscle functions and arterial stiffness appear before puberty in obese children and are associated with elevated ambulatory blood pressure. Eur Heart J 29:792–799
208. Woo KS, Chook P, Yu CW, Sung RY, Qiao M, Leung SS, Lam CW, Metreweli C, Celermajer DS (2004) Overweight in children is associated with arterial endothelial dysfunction and intima-media thickening. Int J Obes Relat Metab Disord 28:852–857
209. Kallio K, Jokinen E, Raitakari OT, Hamalainen M, Siltala M, Volanen I, Kaitosaari T, Viikari J, Ronnemaa T, Simell O (2007) Tobacco smoke exposure is associated with attenuated endothelial function in 11-year-old healthy children. Circulation 115:3205–3212
210. Leeson CP, Kattenhorn M, Morley R, Lucas A, Deanfield JE (2001) Impact of low birth weight and cardiovascular risk factors on endothelial function in early adult life. Circulation 103:1264–1268
211. Leeson CP, Whincup PH, Cook DG, Donald AE, Papacosta OLA, Deanfield JE (1997) Flow-mediated dilation in 9- to 11-year-old children: the influence of intrauterine and childhood factors. Circulation 96:2223–2238
212. Slyper AH (2004) Clinical review 168: what vascular ultrasound testing has revealed about pediatric atherogenesis, and a potential clinical role for ultrasound in pediatric risk assessment. J Clin Endocrinol Metab 89:3089–3095
213. Abbott RA, Harkness MA, Davies PS (2002) Correlation of habitual physical activity levels with flow-mediated dilation of the brachial artery in 5–10 year old children. Atherosclerosis 160:233–239
214. Kelly AS, Wetzsteon RJ, Kaiser DR, Steinberger J, Bank AJ, Dengel DR (2004) Inflammation, insulin, and endothelial function in overweight children and adolescents: the role of exercise. J Pediatr 145:731–736
215. Watts K, Beye P, Siafarikas A, O'Driscoll G, Jones TW, Davis EA, Green DJ (2004) Effects of exercise training on vascular function in obese children. J Pediat 144:620–625
216. Watts K, Beye P, Siafarikas A, Davis EA, Jones TW, O'Driscoll G, Green DJ (2004) Exercise training normalizes vascular dysfunction and improves central adiposity in obese adolescents. J Am Coll Cardiol 43:1823–1827
217. Wiltshire EJ, Gent R, Hirte C, Pena A, Thomas DW, Couper JJ (2002) Endothelial dysfunction relates to folate status in children and adolescents with type 1 diabetes. Diabetes 51:2282–2286

218. Bennett-Richards K, Kattenhorn M, Donald A, Oakley G, Varghese Z, Rees L, Deanfield JE (2002) Does oral folic acid lower total homocysteine levels and improve endothelial function in children with chronic renal failure? Circulation 105:1810–1815
219. Bennett-Richards KJ, Kattenhorn M, Donald AE, Oakley GR, Varghese Z, Bruckdorfer KR, Deanfield JE, Rees L (2002) Oral L-arginine does not improve endothelial dysfunction in children with chronic renal failure. Kidney Int 62:1372–1378
220. Engler MM, Engler MB, Malloy MJ, Chiu EY, Schloetter MC, Paul SM, Stuehlinger M, Lin KY, Cooke JP, Morrow JD, Ridker PM, Rifai N, Miller E, Witztum JL, Mietus-Snyder M (2003) Antioxidant vitamins C and E improve endothelial function in children with hyperlipidemia: Endothelial Assessment of Risk from Lipids in Youth (EARLY) trial. Circulation 108:1059–1063
221. Engler MM, Engler MB, Malloy M, Chiu E, Besio D, Paul S, Stuehlinger M, Morrow J, Ridker P, Rifai N, Mietus-Snyder M (2004) Docosahexaenoic acid restores endothelial function in children with hyperlipidemia: results from the EARLY study. Int J Clin Pharmacol Ther 42:672–679
222. de Jongh S, Vissers MN, Rol P, Bakker HD, Kastelein JJ, Stroes ES (2003) Plant sterols lower LDL cholesterol without improving endothelial function in prepubertal children with familial hypercholesterolaemia. J Inherit Metab Dis 26:343–351
223. Corretti MC, Anderson TJ, Benjamin EJ, Celermajer D, Charbonneau F, Creager MA, Deanfield J, Drexler H, Gerhard-Herman M, Herrington D, Vallance P, Vita J, Vogel R (2002) International brachial artery reactivity task F guidelines for the ultrasound assessment of endothelial-dependent flow-mediated vasodilation of the brachial artery: a report of the International Brachial Artery Reactivity Task Force. J Am Coll Cardiol 39:257–265
224. Aeschlimann SE, Mitchell CK, Korcarz CE (2004) Ultrasound brachial artery reactivity testing: technical considerations. J Am Soc Echocardiogr 17:697–699
225. Tsai WC, Li YH, Lin CC, Chao TH, Chen JH (2004) Effects of oxidative stress on endothelial function after a high-fat meal. Clin Sci 106:315–319
226. Etsuda H, Takase B, Uehata A, Kusano H, Hamabe A, Kuhara R, Akima T, Matsushima Y, Arakawa K, Satomura K, Kurita A, Ohsuzu F (1999) Morning attenuation of endothelium-dependent, flow-mediated dilation in healthy young men: possible connection to morning peak of cardiac events? Clinical Cardiology 22:417–421
227. Gaenzer H, Sturm W, Kirchmair R, Neumayr G, Ritsch A, Patsch J (2000) Circadian variation of endothelium-dependent vasodilatation of the brachial artery as a confounding factor in the evaluation of endothelial function. Atherosclerosis 149:227–228
228. English JL, Jacobs LO, Green G, Andrews TC (1998) Effect of the menstrual cycle on endothelium-dependent vasodilation of the brachial artery in normal young women. Am J Cardiol 82:256–258
229. Lambert J, Stehouwer CD (1996) Modulation of endothelium-dependent, flow-mediated dilatation of the brachial artery by sex and menstrual cycle. Circulation 94:2319–2320
230. Schroeder S, Enderle MD, Baumbach A, Ossen R, Herdeg C, Kuettner A, Karsch KR (2000) Influence of vessel size, age and body mass index on the flow-mediated dilatation (FMD%) of the brachial artery. Int J Cardiol 76:219–225
231. Silber HA, Ouyang P, Bluemke DA, Gupta SN, Foo TK, Lima JA (2005) Why is flow-mediated dilation dependent on arterial size? Assessment of the shear stimulus using phase-contrast magnetic resonance imaging. Am J Physiol Heart Circ Physiol 288:H822–H828
232. Jarvisalo MJ, Ronnemaa T, Volanen I, Kaitosaari T, Kallio K, Hartiala JJ, Irjala K, Viikari JS, Simell O, Raitakari OT (2002) Brachial artery dilatation responses in healthy children and adolescents. Am J Physiol Heart Circ Physiol 282:H87–H92
233. Uehata A, Lieberman EH, Gerhard MD, Anderson TJ, Ganz P, Polak JF, Creager MA, Yeung AC (1997) Noninvasive assessment of endothelium-dependent flow-mediated dilation of the brachial artery. Vasc Med 2:87–92
234. Accini JL, Sotomayor A, Trujillo F, Barrera JG, Bautista L, Lopez-Jaramillo P (2001) Colombian study to assess the use of noninvasive determination of endothelium-mediated vasodilatation (CANDEV). Normal values and factors associated. Endothelium 8:157–166

235. Sorensen K, Celermajer D, Spiegelhalter D, Georgakopoulos D, Robinson J, Thomas O, Deanfield J (1995) Non-invasive measurement of human endothelium dependent arterial responses: accuracy and reproducibility. Br Heart J 74:247–253
236. Bonetti PO, Barsness GW, Keelan PC, Schnell TI, Pumper GM, Kuvin JT, Schnall RP, Holmes DR, Higano ST, Lerman A (2003) Enhanced external counterpulsation improves endothelial function in patients with symptomatic coronary artery disease. J Am Coll Cardiol 41:1761–1768
237. Kuvin J, Patel A, Sliney K, Pandian N, Sheffy J, Schnall R, Karas R, Udelson J (2003) Assessment of peripheral vascular endothelial function with finger arterial pulse wave amplitude. Am Heart J 146:168–174
238. Hamburg NM, Keyes MJ, Larson MG, Vasan RS, Schnabel R, Pryde MM, Mitchell GF, Sheffy J, Vita JA, Benjamin EJ (2008) Cross-sectional relations of digital vascular function to cardiovascular risk factors in the Framingham Heart Study. Circulation 117:2467–2474
239. O'Brien LM, Holbrook CR, Vanderlaan M, Amiel J, Gozal D (2005) Autonomic function in children with congenital central hypoventilation syndrome and their families. Chest 128:2478–2484
240. Tauman R, O'Brien LM, Mast BT, Holbrook CR, Gozal D (2004) Peripheral arterial tonometry events and electroencephalographic arousals in children. Sleep 27:502–506
241. Haller MJ, Stein J, Shuster J, Theriaque D, Silverstein J, Schatz DA, Earing MG, Lerman A, Mahmud FH (2007) Peripheral artery tonometry demonstrates altered endothelial function in children with type 1 diabetes. Pediatr Diabetes 8:193–198
242. Mahmud FH, Van Uum S, Kanji N, Thiessen-Philbrook H, Clarson CL, Mahmud FH, Van Uum S, Kanji N, Thiessen-Philbrook H, Clarson CL (2008) Impaired endothelial function in adolescents with type 1 diabetes mellitus. J Pediatr 152:557–562
243. Cracowski JL, Minson CT, Salvat-Melis M, Halliwill JR (2006) Methodological issues in the assessment of skin microvascular endothelial function in humans. Trends Pharmacol Sci 27:503–508
244. Vuilleumier P, Decosterd D, Maillard M, Burnier M, Hayoz D (2002) Postischemic forearm skin reactive hyperemia is related to cardovascular risk factors in a healthy female population. [see comment]. J Hypertens 20:1753–1757
245. Elhadd TA, Khan F, Kirk G, McLaren M, Newton RW, Greene SA, Belch JJ (1998) Influence of puberty on endothelial dysfunction and oxidative stress in young patients with type 1 diabetes. Diabetes Care 21:1990–1996
246. Khan F, Elhadd TA, Greene SA, Belch JJ (2000) Impaired skin microvascular function in children, adolescents, and young adults with type 1 diabetes. Diabetes Care 23:215–220
247. Gozal D, Kheirandish-Gozal L, Serpero LD, Sans Capdevila O, Dayyat E (2007) Obstructive sleep apnea and endothelial function in school-aged nonobese children: effect of adenotonsillectomy. Circulation 116:2307–2314
248. Gaeta G, De Michele M, Cuomo S, Guarini P, Foglia MC, Bond MG, Trevisan M (2000) Arterial abnormalities in the offspring of patients with premature myocardial infarction. N Engl J Med 343:840–846
249. Singhal A, Cole TJ, Fewtrell M, Deanfield J, Lucas A (2004) Is slower early growth beneficial for long-term cardiovascular health? Circulation 109:1108–1113
250. Fernandez-Real JM, Castro A, Vazquez G, Casamitjana R, Lopez-Bermejo A, Penarroja G, Ricart W (2004) Adiponectin is associated with vascular function independent of insulin sensitivity. Diabetes Care 27:739–745
251. Olsen MH, Wachtell K, Hermann KL, Bella JN, Dige-Petersen H, Rokkedal J, Ibsen H (2002) Left ventricular hypertrophy is associated with reduced vasodilatory capacity in the brachial artery in patients with longstanding hypertension. A LIFE substudy. Blood Press 11:285–292
252. Papaioannou GI, Seip RL, Grey NJ, Katten D, Taylor A, Inzucchi SE, Young LH, Chyun DA, Davey JA, Wackers FJ, Iskandrian AE, Ratner RE, Robinson EC, Carolan S, Engel S, Heller GV (2004) Brachial artery reactivity in asymptomatic patients with type 2 diabetes

mellitus and microalbuminuria (from the Detection of Ischemia in Asymptomatic Diabetics-brachial artery reactivity study). Am J Cardiol 94:294–299
253. Chambers JC, Obeid OA, Kooner JS (1999) Physiological increments in plasma homocysteine induce vascular endothelial dysfunction in normal human subjects. Arterioscler Thromb Vasc Biol 19:2922–2927
254. Chenevard R, Hurlimann D, Bechir M, Enseleit F, Spieker L, Hermann M, Riesen W, Gay S, Gay RE, Neidhart M, Michel B, Luscher TF, Noll G, Ruschitzka F (2003) Selective COX-2 inhibition improves endothelial function in coronary artery disease. Circulation 107:405–409

Chapter 10
The Cardiovascular Risk in Young Finns Study and the Special Turku Coronary Risk Factor Intervention Project (STRIP)

Markus Juonala, Costan G. Magnussen, Olli Simell, Harri Niinikoski, Olli T. Raitakari and Jorma S. A. Viikari

Abstract In the longitudinal Young Finns study, cardiovascular risk factors were first assessed in 1980 in 3,596 children aged 3–18 years. The latest follow-up examinations were performed in 2001 and 2007, when risk factors and early markers of atherosclerosis in carotid and brachial arteries were examined among 2,653 subjects from the original cohorts, now aged 24–45 years. The results show that an individual's risk factor profile is regulated by early lifestyle-related factors and that exposure to risk factors in childhood induces changes in arteries that contribute to the development of atherosclerosis in adulthood. In the STRIP study, 1,062 infants were randomized into an intervention group (n=540; low-saturated-fat, low-cholesterol diet) or a control group (n=522) at 7 mo of age in 1989. STRIP has demonstrated the benefits, e.g. improved lipid levels and brachial endothelial function, of low-saturated-fat diet that could be initiated at an early age. In addition, active exercise habits and non-smoking should be encouraged since they influence vascular endothelial function in childhood. No stronger justification for aiming at improvement of CVD risk factor levels, including serum cholesterol values, blood pressure and obesity, can be provided.

Keywords Atherosclerosis • Blood pressure • Children • Cholesterol • Coronary heart disease • Diet • Flow mediated dilatation • Intervention • Intima-media thickness • Lipids • Prevention • Risk factors

10.1 The Cardiovascular Risk in Young Finns Study

In Finland, coronary heart disease (CHD) incidence was very high in the 1960s and 1970s. In line with this, the Seven Countries Study showed that the level of serum cholesterol in Finns was also the highest among the investigated countries in the 1960s [1].

M. Juonala (✉)
Department of Pediatrics, Medicine and Clinical Physiology, Research Center of Applied and Preventive Cardiovascular Medicine, University of Turku and Turku University Hospital, Turku, Finland
e-mail: mataju@utu.fi

G. S. Berenson (ed.), *Evolution of Cardio-Metabolic Risk from Birth to Middle Age*,
DOI 10.1007/978-94-007-1451-9_10, © Springer Science+Business Media B.V. 2011

With reference to the studies indicating that atherosclerosis starts early [2], the World Health Organization issued a recommendation in 1978 that epidemiological studies investigating childhood cardiovascular risk should be initiated. Inspired by those facts, as well as results of the Bogalusa Heart Study [3, 4], a program was launched in Finland in the late 1970s to study cardiovascular risk in youth [5]. The Cardiovascular Risk in Young Finns Study was designed as a collaborative effort between five university departments with medical schools (i.e. in Helsinki, Kuopio, Oulu, Tampere and Turku) and several other institutions in Finland. The aim was to study the levels of CHD risk factors and their determinants in children and adolescents of various ages in different parts of the country. The first cross-sectional study was performed in 1980. The baseline study included 3,596 children and adolescents aged 3, 6, 9, 12, 15 and 18 years [5, 6]. Between 1980 and 1992, these cohorts were followed-up in 3-year intervals. The 21-year and 27-year follow-ups were performed in 2001 and 2007 when data on 2,284 and 2,204 of the original participants, now young adults, was collected. In addition, the 30-year follow-up is starting in late 2010 (Fig. 10.1).

The main aim of the Young Finns Study has been to examine whether childhood risk factor levels are good indicators of cardiovascular health in adulthood. In the absence of sufficient event data, carotid intima-media thickness (IMT), carotid elasticity and brachial artery flow-mediated dilation (FMD) measured in the 21- and 27-year follow-ups have been used as surrogate markers. By far, the findings have supported the hypothesis that exposure to risk factors in childhood influence atherosclerosis development. For example, increased levels of LDL-cholesterol,

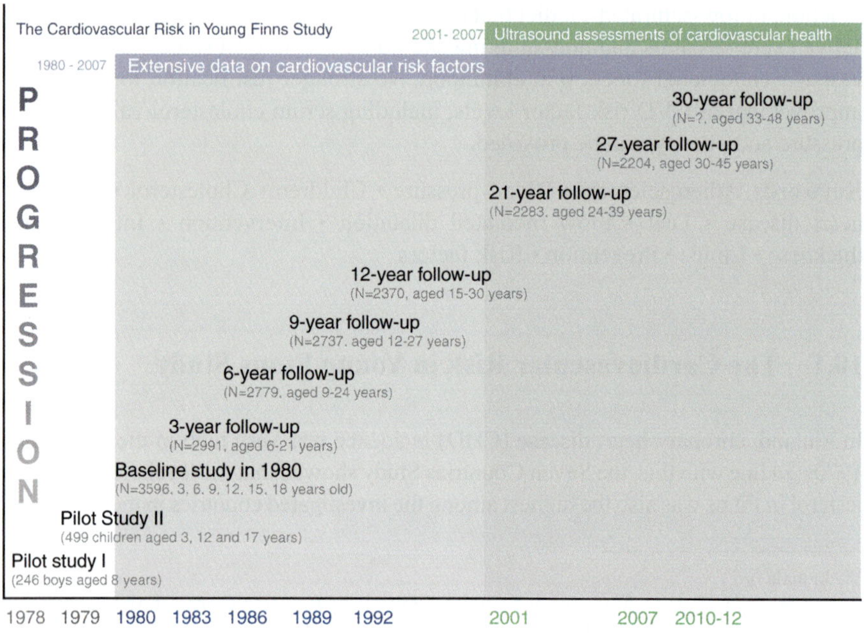

Fig. 10.1 The progression of the Young Finns Study

systolic blood pressure, pulse pressure, BMI, as well as cigarette smoking in those aged 12–18 years predicted increased IMT in adulthood [7]; systolic blood pressure measured at ages 12–18 years was inversely associated with endothelial function in adulthood independent of other adolescent and adult risk factors [8]; and systolic blood pressure and skinfold thickness were inversely associated with arterial elasticity at follow-up in the entire cohort [9]. Two extremely important observations from these reports have been that the extent of preclinical atherosclerosis increases as the number of childhood risk factors increases, and that risk factors measured in childhood appear to be stronger predictors of preclinical atherosclerosis than measurements obtained at the time imaging was performed [7, 9]. Together, these findings provide a strong rationale for both the identification of children at high-risk and for efforts to prevent the development of CVD risk factors in early life [10]. In addition, childhood risk factors, especially life-style, such as diet and physical activity, have been shown to predict increased IMT progression rate in young adults [11].

We have shown that the presence of the metabolic syndrome, the clustering of several interrelated cardiometabolic risk factors, in adulthood to be contingent on childhood obesity, hypertriglyceridemia, hyperinsulinemia, high CRP, and family history of hypertension and type 2 diabetes [12]. It has also been observed that metabolic syndrome is independently associated both cross-sectional carotid IMT and IMT progression rate among young adults [13, 14]. However, we have shown that recovery from the metabolic syndrome over a 6-year follow-up in adulthood to be associated with favorable changes in carotid artery structure and function [15], which is an important finding from a public health perspective.

The main results from the Young Finns Study are outlined in Table 10.1. Most importantly, the findings have clearly shown that an individual's cardiovascular risk profile in adulthood is significantly modified by childhood risk factors and lifestyle-related factors. Therefore, maintaining optimal weight, adequate levels of physical activity, non-smoking and prudent diet with frequent fruit consumption from childhood to adulthood should be encouraged.

Table 10.1 Main findings of the Young Finns and STRIP studies

The cardiovascular risk in Young Finns Study

- Childhood risk factors predict adult carotid IMT and elasticity [7, 9]
- Childhood physical activity and diet associate with IMT progression in adults [11]
- Excellent endothelial function seems to protect from adverse risk factor effects on IMT [32]
- Metabolic syndrome is associated with early atherosclerosis and its progression [13, 14]
- Recovery from metabolic syndrome restores subclinical changes in carotid artery [15]

The STRIP study

- Dietary intervention initiated in infancy improves LDL-cholesterol levels [16, 18]
- Intervention is safe in terms of growth, pubertal development and childhood neurological development [19]
- Among boys, intervention group has improved endothelial function [20]
- Physical activity is associated with enhanced endothelial function [21]
- Exposure to tobacco smoke is related with impaired brachial FMD [22]

In the near future, the Young Finns Study aims to extend beyond investigating the effects of conventional cardiovascular risk factors to explore the influence of genetic variation on cardiovascular risk. Genome-wide analysis has just been completed which has allowed ~600,000 single-nucleotide polymorphisms to be determined amongst the study participants, providing a unique opportunity to examine gene, gene–gene, and gene–environment findings. In addition, to gain more data on the effects of childhood risk factors on adult health, cardiac and liver ultrasound studies and cognitive function testing will be performed during the 30-year follow-up (Fig. 10.1).

10.2 The Special Turku Coronary Risk Factor Intervention Project (STRIP)

The STRIP study was initiated in 1990 [16, 17]. It is a randomized intervention trial, where a low-saturated-fat, low-cholesterol diet and lifestyle counselling was started in healthy infants in the intervention group before their first birthday. Since then, 540 intervention children and 522 controls have been prospectively followed every 6–12 months throughout childhood and adolescence. The study subjects are currently 18–20 years old and >50% of the initial study children are still participating in the project (Fig. 10.2).

Dietary counselling of the intervention families was individualized and aimed at child's fat intake of 30–35% of daily energy (E%), saturated to monounsaturated+polyunsaturated fatty acid ratio (S/M+P) of 1:2 and cholesterol intake <200 mg/day. Breast milk or formula was advised during the first year of life and

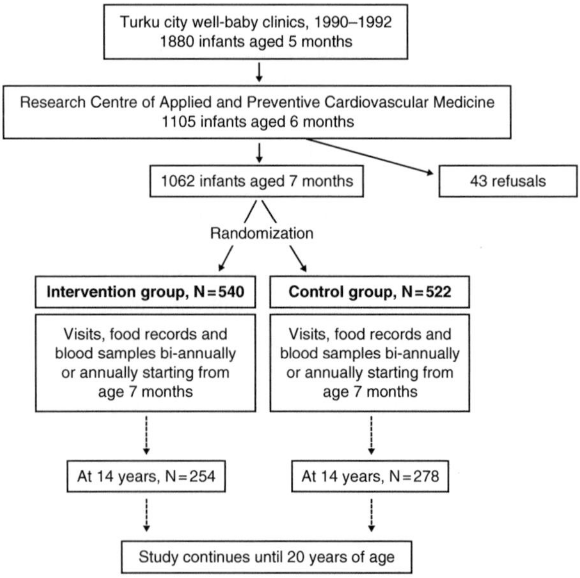

Fig. 10.2 The progression of the STRIP study

thereafter 0.5–0.6 L skim milk daily was recommended. To maintain adequate fat intake, the parents were instructed to add two to three teaspoons (approx. 10 g) of soft margarine or vegetable oil (mainly low-erucic-acid rapeseed oil) daily to the child's food from 12 to 24 months of age. This fundamentally changed the dietary fat *quality* without modifying the amount of fat in diet. The families were encouraged to gradually change their child's diet towards a better fat composition by, e.g., preferring vegetable oil over animal fat. Ample use of vegetables, fruits, berries and whole grain products was encouraged. The counselling was always based on the age and cognitive ability of the child, and from age 7 years onwards progressively more dietary information and suggestions were given directly to the child. Control children received basic health education routinely provided at Finnish well-baby clinics and school health care. According to the general recommendations in Finland, at the age of 12 months cow's milk with 1.9% (1.5% after May 1995) fat was recommended to the control children for daily use. Dietary issues were discussed only superficially with control families.

The main objectives of STRIP have been to study the safety and effects of low-saturated-fat diet on atherosclerosis risk factors in childhood. This is important since fears that low intake of saturated fat and cholesterol might influence growth and cognitive development have led to exclusion of infants and young children from dietary fat quantity and quality modification recommendations in many countries. Fat is an important source of energy, especially in infancy, when up to 50% of daily energy intake of breast-fed infants comes from fat. A diet with too low fat content might, due to low energy density, be harmful to growth. However, the goal in STRIP is to change the *quality* of fat rather than *quantity* of it. Change in fat quality is, indeed, a more effective way of improving serum cholesterol and lipoprotein values in children than change in its amount.

The results of the STRIP study have been encouraging. The intervention children have had lower fat and saturated fat intakes and higher protein and carbohydrate intakes than control children [18]. The energy intake of the intervention children, especially of the boys, was also slightly lower than that of the control children throughout the study. The dietary intervention significantly decreased serum cholesterol values [18] throughout childhood and this ~5% difference between intervention and control children persisted at least through the first 14 years of life in boys. Importantly, cognitive tests were performed at the age of 5 years to determine if the intervention influences speech and language or visual motor skills or gross motor functioning, but no negative effects were found [19]. Overall, the intervention effect has been stronger in boys than in girls, but in the latest follow-up at the age of 17 years a significant LDL-C lowering effect was observed also in girls (unpublished observations). Serum triglyceride values were lower in boys than in girls, and intervention had an effect on serum triglyceride values in boys. Adjustment with energy intake, saturated fat intake (as E%), height (SD) and BMI did not alter the results.

In STRIP children, vascular endothelial function has been assessed at 11 years of age and biannually thereafter. FMD was enhanced in 11-year-old intervention boys compared with control boys [20]. Interestingly, while the mean serum cholesterol concentration measured during the life course correlated with brachial FMD in both sexes, the cholesterol concentration measured before the age of 5 years was a stronger correlate than mean cholesterol concentration measured after this age. Furthermore, the difference in endothelial function between intervention boys and control boys remained significant after adjustment for current LDL-cholesterol concentrations. However, the difference became non-significant after taking into account the mean cholesterol concentration measured between ages 7 months and 2 years. These observations were interpreted to suggest that the higher FMD responses seen in boys in the intervention group was not merely a reflection of recent cholesterol control, but that it reflects the influence of long-term cholesterol control in vascular function.

The STRIP population has also been treated as a unique cohort study to gain insights on the other determinants of early vascular changes in children. In these analyses, leisure-time physical activity has been shown to associate with brachial FMD in boys [21], while FMD has been shown to be enhanced in children never exposed to tobacco smoke compared with those with constant passive smoke exposure [22].

STRIP has demonstrated the benefits of low-saturated-fat diet that could be initiated at an early age (Table 10.1). This is best achieved by using 0% fat milk (with diet supplemented with vegetable oil, e.g. low-erucic acid rapeseed oil, to ensure sufficient fat intake) and other low-saturated-fat dairy products and meat after 1 year of age. Moreover, to improve fat quality, vegetable oil should be used instead of butter. Active exercise habits and non-smoking should be encouraged since they influence vascular endothelial function in childhood. No stronger justification for aiming at improvement of CVD risk factor levels, including serum cholesterol values, blood pressure and obesity, can be provided [10].

10.3 The International Childhood Cardiovascular Cohort (i3C) Consortium

The Young Finns Study is also an integral part of an international consortium of prospective cohort studies beginning in childhood (i3C), Bogalusa, Muscatine, Finland, and Australia [23–26]. By pooling event data across the cohorts, it is expected that analyses linking childhood risk factors to CVD morbidity and mortality outcomes will be completed sooner. While these analyses are still some time off, pooled analyses performed as part of the consortium have shown pediatric dyslipidemia levels to predict adult dyslipidemia [23] and high carotid IMT [24]. Pediatric metabolic syndrome has been observed to predict type 2 diabetes and subclinical atherosclerosis in adulthood [25]. It has also been found that the strength of the associations between childhood risk factors and carotid IMT are dependent on childhood age. Based on the consortium data, risk factor measurements obtained at or after 9 years of age are predictive of subclinical atherosclerosis in adulthood [26].

10.4 Lifelong, Primordial Prevention of Cardiovascular Disease is Important

In line with the Young Finns Study results (Table 10.1), observations from the Bogalusa Heart Study [27] and the Muscatine Study [28] have outlined the importance of childhood risk factors in determining carotid IMT in adults. Most importantly, the landmark articles from the Bogalusa Heart Study [29, 30] have shown that childhood risk factors are predictive for atherosclerotic lesions in the aorta and coronary arteries. These findings have given strong support for the concept of primordial prevention in which the main goal is to improve risk factor profile in childhood and thereby decrease the lifelong risk of cardiovascular disease [31], instead of starting primary and secondary prevention among adults already suffering from clinical cardiovascular disease or adverse risk factor burden. In the sense of pursuing this lifelong approach, the results from the STRIP study (Table 10.1) have provided feasible means for this preventive work.

Reference

1. Keys A, Aravanis C, Blackburn HW et al (1996) Epidemiological studies related to coronary heart disease: characteristics of men aged 40–59 in seven countries. Acta Med Scand Suppl 460:1–392
2. Enos WF, Holmes RH, Beyer J (1953) Coronary disease among United States soldiers killed in action in Korea: preliminary report. JAMA 152:1090–1093
3. Srinivasan SR, Frerichs RR, Berenson GS (1978) Serum lipids and lipoproteins in children. In: Strong WB (ed) Atherosclerosis: its pediatric aspects. Grune & Stratton, New York, 85–110
4. Berenson GS (1985) Epidemiologic investigations of cardiovascular risk factor variables in childhood—an overview. Acta Paediatr Scand Suppl 318:7–9
5. Åkerblom HK, Viikari J, Uhari M et al (1985) Atherosclerosis precursors in Finnish children and adolescents. I. General description of the cross-sectional study of 1980, and an account of the children's and families' state of health. Acta Paediatr Scand Suppl 318:49–63
6. Raitakari OT, Juonala M, Rönnemaa T et al (2008) Cohort profile: the cardiovascular risk in Young Finns Study. Int J Epidemiol 37:1220–1226
7. Raitakari OT, Juonala M, Kähönen M et al (2003) Cardiovascular risk factors in childhood and carotid artery intima-media thickness in adulthood—the cardiovascular risk in Young Finns Study. JAMA 290:2277–2283
8. Juonala M, Viikari JSA, Rönnemaa T et al (2006) Elevated blood pressure in adolescent boys predicts endothelial dysfunction: the cardiovascular risk in Young Finns Study. Hypertension 48:424–430
9. Juonala M, Järvisalo MJ, Mäki-Torkko N et al (2005) Risk factors identified in childhood and decreased carotid artery elasticity in adulthood. The cardiovascular eisk in Young Finns Study. Circulation 112:1489–1496
10. Gidding SS (2006) New cholesterol guidelines for children? Circulation 114:989–991
11. Juonala M, Viikari JSA, Kähönen M et al (2010) Life-time risk factors and progression of carotid atherosclerosis in young adults. The cardiovascular risk in Young Finns Study. Eur Heart J 31:1745–1751

12. Mattsson N, Rönnemaa T, Juonala M et al (2008) Childhood predictors of the metabolic syndrome in adulthood. The cardiovascular risk in Young Finns Study. Ann Med 40:542–552
13. Mattsson N, Rönnemaa T, Juonala M et al (2008) Arterial structure and function in young adults with the metabolic syndrome: the cardiovascular risk in Young Finns Study. Eur Heart J 29:784–791
14. Koskinen J, Kähönen M, Viikari JS et al (2009) Conventional cardiovascular risk factors and metabolic syndrome in predicting carotid intima-media thickness progression in young adults: the cardiovascular risk in Young Finns Study. Circulation 120:229–236
15. Koskinen J, Magnussen CG, Taittonen L et al (2010) Arterial structure and function after recovery from the metabolic syndrome: the cardiovascular risk in Young Finns Study. Circulation 121:392–400
16. Lapinleimu H, Viikari J, Jokinen E et al (1995) Prospective randomized trial in 1062 infants of diet low in saturated fat and cholesterol. Lancet 345:471–476
17. Simell O, Niinikoski H, Rönnemaa T et al (2008) Cohort profile: the STRIP study (Special Turku Coronary Risk Factor Intervention Project), an infancy-onset dietary and life-style intervention trial. Int J Epidemiol 38:650–655
18. Niinikoski H, Lagström H, Jokinen E et al (2007) Impact of repeated dietary counseling between infancy and 14 years of age on dietary intakes and serum lipids and lipoproteins: the STRIP study. Circulation 116:1032–1040
19. Rask-Nissila L, Jokinen E, Terho P et al (2000) Neurological development of 5-year-old children receiving a low-saturated fat, low-cholesterol diet since infancy—a randomized controlled trial. JAMA 284:993–1000
20. Raitakari OT, Rönnemaa T, Järvisalo MJ et al (2005) Endothelial function in healthy 11-year-old children after dietary intervention with onset in infancy: the Special Turku Coronary Risk Factor Intervention Project for children (STRIP). Circulation 112:3786–3794
21. Pahkala K, Heinonen OJ, Lagström H et al (2008) Vascular endothelial function and leisure-time physical activity in adolescents. Circulation 118:2353–2359
22. Kallio K, Jokinen E, Raitakari OT et al (2007) Tobacco smoke exposure is associated with attenuated endothelial function in 11-year-old healthy children. Circulation 115:3205–3212
23. Magnussen CG, Raitakari OT, Thomson R et al (2008) Utility of currently recommended pediatric dyslipidemia classifications in predicting dyslipidemia in adulthood: evidence from the Childhood Determinants of Adult Health (CDAH) study, cardiovascular risk in Young Finns Study, and Bogalusa Heart Study. Circulation 117:32–42
24. Magnussen CG, Venn A, Thomson R et al (2009) The association of pediatric LDL-cholesterol and HDL-cholesterol dyslipidemia classifications and change in dyslipidemia status with carotid intima-media thickness in adulthood: evidence from the cardiovascular risk in Young Finns Study, the Bogalusa Heart Study, and the Childhood Determinants of Adult Health (CDAH) study. JACC 53:860–869
25. Magnussen CG, Koskinen J, Chen W et al (2010) Pediatric metabolic syndrome predicts adulthood metabolic syndrome, subclinical atherosclerosis, and type 2 diabetes mellitus but is no better than body mass index alone: the Bogalusa Heart Study and the cardiovascular risk in Young Finns Study. Circulation 122:1604–1611
26. Juonala M, Magnussen CG, Venn A et al (2010) The influence of age on associations between childhood risk factors and carotid intima-media thickness in adulthood. The cardiovascular risk in Young Finns Study, the Childhood Determinants of Adult Health Study, the Bogalusa Heart Study and the Muscatine Study for the International Childhood Cardiovascular Cohort (i3C) Consortium. Circulation 122:2514–2520
27. Li S, Chen W, Srinivasan SR et al (2003) Childhood cardiovascular risk factors and carotid vascular changes in adulthood: the Bogalusa Heart Study. JAMA 290:2271–2276
28. Davis PH, Dawson JD, Riley WA et al (2001) Carotid intimal-medial thickness is related to cardiovascular risk factors measured from childhood through middle age: the Muscatine Study. Circulation 104:2815–2819

29. Newman WP, III, Freedman DS, Voors AW et al (1986) Relation of serum lipoprotein levels and systolic blood pressure to early atherosclerosis. The Bogalusa Heart Study. N Engl J Med 314:138–144
30. Berenson GS, Srinivasan SR, Bao W et al (1998) Association between multiple cardiovascular risk factors and atherosclerosis in children and young adults. The Bogalusa Heart Study. N Engl J Med 338:1650–1656
31. Capewell S, Lloyd-Jones DM (2010) Optimal cardiovascular prevention strategies for the 21st century. JAMA 304:2057–2058
32. Juonala M, Viikari JSA, Laitinen T et al (2004) Interrelations between brachial endothelial function and carotid intima-media thickness in young adults. The cardiovascular risk in Young Finns Study. Circulation 110:2918–2923

Chapter 11
Prevention of Heart Disease in Childhood—Encouragement of Primordial Prevention

Gerald S. Berenson and Arthur Pickoff

Abstract Observations over the many years in the Bogalusa Heart Study, have established that adverse cardiovascular risk factors have their onset early in life. The most dramatic evidence of cardiovascular disease comes from autopsy studies on young individuals in Bogalusa. Such findings have been extended by the Pathologic Determinants of Atherosclerosis in Youth and very strong correlations occur at autopsy between ante mortem clinical cardiovascular risk factors and actual anatomic changes consistent with cardiovascular disease. Lifestyles that are associated with atherosclerosis and hypertension—high fat-high cholesterol diets, high salt intake, inactivity, obesity, consistent smoking (we noted by 8–9 years of age) and other poor lifestyles need to be addressed. These findings urge the medical profession, especially cardiologists, to give leadership to prevention involving children and their families. Addressing lifestyles and behavior for families including their offspring should be a high priority to break this devastating toll on our society. Although there are now many drugs available that improve medical aspects and even prolong life by treating coronary artery disease, hypertension and diabetes mellitus, improving poor lifestyles beginning at a young age could work in concert with both primary and secondary prevention. Yet, the ubiquitous prevalence of cardiovascular diseases and their beginning at an early age, show the need to improve deleterious lifestyles for all children and begin primordial prevention. Primordial prevention through public health education at the elementary school age with family involvement should be a goal for Preventive Cardiology.

Keywords Prevention • Primordial prevention • Lifestyles • Health education • Public education • Elementary school children • Role models • Preventive Cardiology

G. S. Berenson (✉)
Department of Medicine, Pediatrics, Biochemistry, Epidemiology, Center for Cardiovascular Health, Tulane University School of Medicine and School of Public Health and Tropical Medicine, New Orleans, LA, USA
e-mail: berenson@tulane.edu

G. S. Berenson (ed.), *Evolution of Cardio-Metabolic Risk from Birth to Middle Age*,
DOI 10.1007/978-94-007-1451-9_11, © Springer Science+Business Media B.V. 2011

11.1 Introduction

Prevention or delay onset of heart disease and to improve quality of life are fundamental priorities of all the research conducted in the Bogalusa Heart Study [1–3]. The fact that adult heart diseases begin in childhood has been clearly established by multiple long term pediatric studies. The most detailed and long term are the Muscatine Study [4], Finnish Youth Study [5], and the Bogalusa Heart Study [1–3]. These now go back almost 40 years. To establish the early onset of cardiovascular disease was an objective of early studies from Bogalusa. Attention must now turn to prevention. Although clinical events related to heart and cardiovascular diseases occur later in life, evidence of subclinical, asymptomatic, and "silent" atherosclerosis, hypertension, and diabetes mellitus is clearly present in young individuals. Testimony to the observation of early onset of heart and cardiovascular diseases has been spelled out in two previous books and the preceding chapters from updated research in the Bogalusa Heart Study. The precise initiating factors are still being investigated but by understanding the earliest determinants allow us to achieve the greatest hope for prevention of adult heart disease.

In the United States mortality from cardiovascular diseases is still the leading cause of death and this is now occurring worldwide, surpassing malnutrition and infectious diseases [6]. The peak of mortality from cardiovascular diseases in the United States occurred around 1965 and with considerable effort of the American Heart Association and programs as an outgrowth of work by the National Institute of Health and Center for Disease Control and Prevention, a rather drastic decrease has occurred. The dramatic onset of obesity over the past three decades has threatened to reverse this decline [7].

We reflect on the underlying determinants of cardiovascular diseases being governed by genetic and environmental interactions. The epidemic of obesity, over the past two decades, has accelerated and by continuing has helped us focus on environment, lifestyles and behaviors. It is of importance that this epidemic has occurred in the Bogalusa population with the same genetic pool. Although selective ethnic tendencies toward obesity and other risk factors occur, this observation emphasizes the need to better understand mechanisms leading to obesity and ethnic risk factors and approaches for prevention.

The early onset of cardiovascular traits, coronary heart disease, atherosclerosis, essential hypertension and diabetes is complex and involves multiple genes being modulated by lifestyle-related behaviors. The tremendous prevalence of risk factors in early life and the tendency for their increase, like obesity, emphasizes that the current efforts for prevention must focus on environment and the related unhealthy lifestyles—oversaturation of high fat, high carbohydrate and excess calories in our diet, sedentary lifestyles of inadequate motion with an imbalance of calories and other behaviors like tobacco use and excess alcohol intake. The physical inactivity beginning in childhood and by young individuals has greatly contributed to the obesity epidemic. But, since a genetic predisposition to the imbalance of energy intake and energy expenditure occurs, there still is a need for further research and refinement of our effort. Despite such a need for more research, we already have the know-how to begin prevention for individuals and broad public application.

11.2 Evidence from Risk Factor Observations for Need of Prevention

Observations of tracking of risk factors from childhood to adulthood provide evidence that risk factors in early life are predictive of adult levels of risk factors, as well as cardiovascular disease [8, 9]. Abnormal levels of risk factors become evident at a young age, Fig. 11.1, while Fig. 11.2, as an example, shows how secular trends of obesity have occurred emphasizing the influences of environment and lifestyle changes. Unfortunately this trend of increasing obesity has continued, Fig. 11.3. Greater inactivity, computer games and television watching of 2–6 h per day, decrease in physical education in schools, and an increase in use of automobiles and buses for transportation of school children have all contributed to less physical activity. This increase in obesity has health implications for greater onset of adult type II diabetes at the adolescent age [8]. Figure 11.4 shows the incidence of hyperinsulinemia as body mass increases. This sets the stage for increased levels of truncal fat and adverse levels and clustering of traditional risk factors, along with increased levels of highly atherogenic, small dense LDL particles, fibrinogen, plasminogen activator inhibitor, and C reactive protein [10].

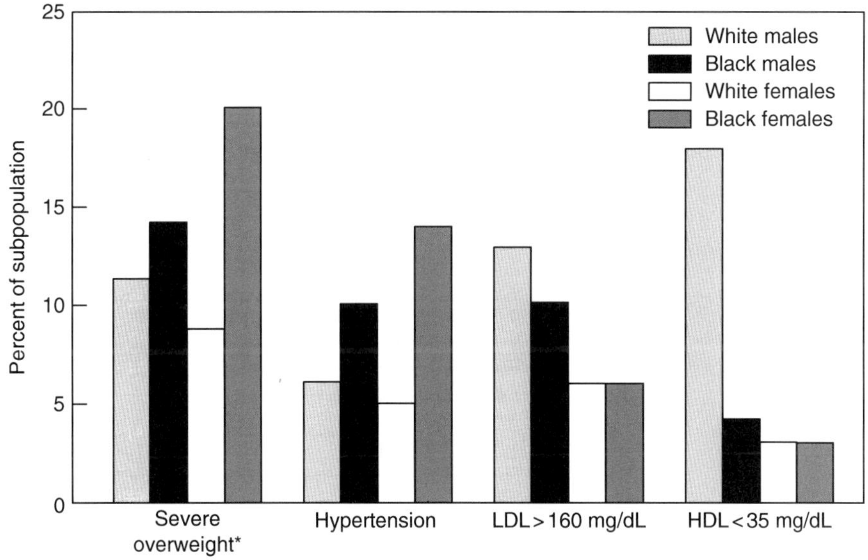

Fig. 11.1 Emergence of clinically abnormal levels of cardiovascular disease risk factors among young adults, ages 19–32 years. [20]
* Body Mass Index > 31.1 kg/m^2 in males and 32.3 kg/m^2 in females

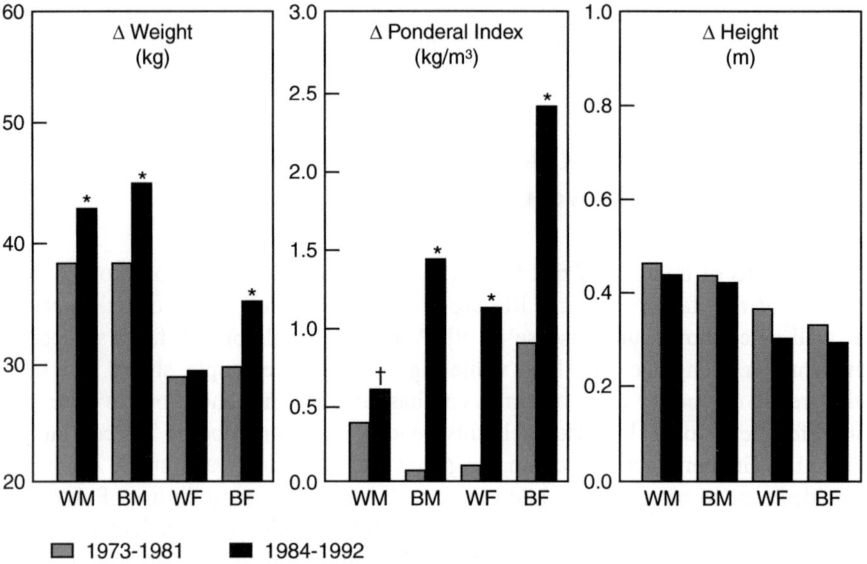

Fig. 11.2 Changes in obesity measures and height in two cohorts, age 7–9 years, over an 8 year period, cohort 1 (1973–1981) and cohort 2 (1984–1992). A trend of increasing obesity is apparent with acceleration of obesity in the more recent years. [20]
WM, white males; BM, black males; WF, white females; BF, black females
*p < 0.01; †p < 0.05

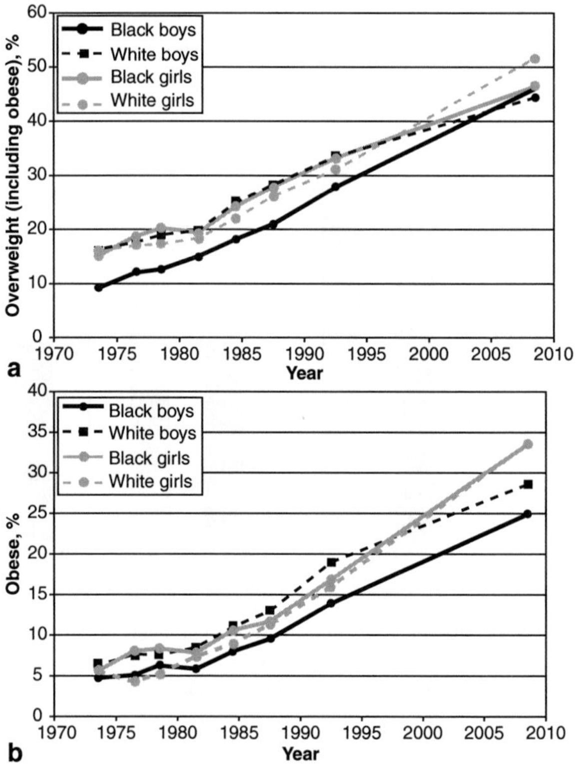

Fig. 11.3 Shows the increasing trend of obesity documented on students in the Bogalusa School System, in 2009. [25]

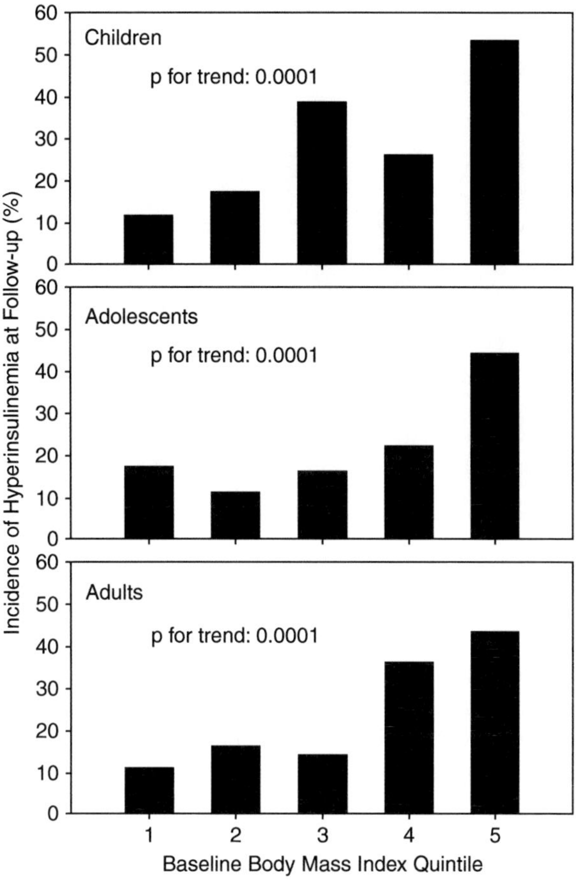

Fig. 11.4 Incidence of hyperinsulinemia (>75th percentile, specific for age, race, sex and survey year) at a 3-year follow-up in children, adolescents and young adults by body mass index quintile (specific for age, race, sex and survey year) at baseline. [26]

Over nutrition or metabolic overload contributes to the major health problem of obesity. Availability of highly palatable, energy dense food for mass consumption along with sedentary lifestyles promotes weight gain. The small amount of excess calories needed for weight gain over time is immeasurable by dietary analytical studies. It is suggested excess 50 kcal per day could result in a 5 lb weight gain over a year.

11.3 Beginning Preventive Cardiology at a Young Age

Preventive cardiology needed at a population level is difficult to accomplish. Figure 11.5 illustrates the complexity of attacking the problem of obesity and addressing other risk factors is equally complex. Primary and secondary prevention have been tremendously expanded with the development of numerous drugs directed at specific risk factors, like hypertension and dyslipidemia. Methods and development of instrumentation for diagnosis and treatment of heart disease patients

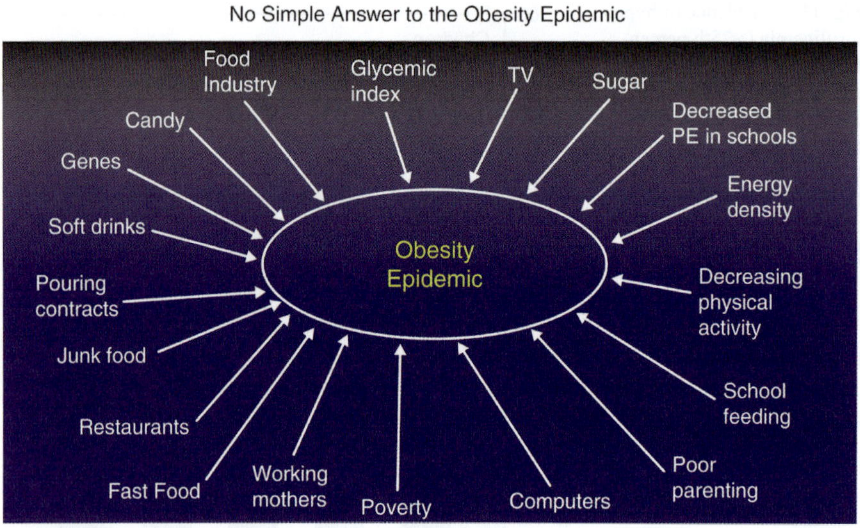

Fig. 11.5 Suggests the complexity of the obesity epidemic

have become part of secondary and primary prevention trials. However, it should be noted that much of this clinical effort involves middle-aged and older individuals. Some notion of including younger individuals comes from the relatively recent diagnoses of "prediabetes" and "prehypertension" [11, 12]. The "pre" is a misnomer since it may lead individuals to think that they do not yet have the disease. However, studies in the Bogalusa Heart Study indicate that a large proportion of individuals with the "pre" diagnosis already have abnormal changes in the cardiovascular system. Elsewhere in this book noninvasive methods adaptable for clinical use are described and have become available to expand characterization of cardiovascular risk beyond the Framingham risk score. The Rasmussen Score is an example of incorporating an instrument measuring vascular compliance with clinical risk factors [13]. Another is the use of percentiles of carotid intima media thickness to provide a "vascular score" [14], and distensibility of the brachial artery has also been used to reflect on changes occurring in the cardiovascular system [15]. These are advances being made at a clinical level that are of tremendous value to determine the impact or burden of risk factors over time on the cardiovascular system. The noninvasive methodologies will provide guidelines for primary prevention once risk factors have been identified.

Much of the findings of the Bogalusa Heart Study have become a basis for prevention beginning early in life. For the most successful outcome from prevention of adult cardiovascular disease prevention has to be "primordial." The concept of primordial prevention was suggested by Tom Strassen from the World Health Organization [16]. This is a logical approach enunciated by Geoffrey Rose who raised the issue of risk in terms of "sick individuals vs. sick populations" [17]. Another symposium at the American Health Foundation held by Ernst Wynder indicated the need to address risk factors in early life [18]. The consideration of primordial prevention is the essence of the Pediatric population risk factor studies.

11.4 Implementing Prevention

We were funded in 1987 as the first National Research and Demonstration Center—Arteriosclerosis to develop practical methods of prevention. Based on the findings of the Bogalusa Heart Study, we were able to develop both a high risk model to control risk factors in families and a public health or population strategy. Table 11.1 shows three models that have been developed with further detail provided by a web page: http://tulane.edu/som/cardiohealth/.

11.4.1 Family Health Promotion

The Family Health Promotion is conducted by a Cardiologist or nurse trained in cardiovascular disease to lead a multidisciplinary team of nutritionist, exercise specialist, and a behavioral-oriented person [19, 20]. The team addresses a group of families together, using a "Weight Watchers" or "Alcohol-Anonymous" group style. Families are selected by having a member with cardiovascular disease or abnormal risk factors. All members are initially screened for cardiovascular risk factors including children. The group is addressed by the team weekly, working together over a period of 10–12 weeks. The physician discusses risk factors with the group and individually as needed. Sessions are provided on diet, food purchasing, menu planning, preparation, and with demonstrations. Physical activity is discussed in detail and appropriate exercises are conducted at each session. Tobacco use, side stream smoke, alcohol use, and their effect on blood pressure and the cardiovascular system are included in the training program. Behavioral "contracts" are signed, for examples, use of skim milk or low fat foods, or no alcohol for one week. Individual counseling is held for special risk factor evaluation and repeated observations are

Table 11.1 Prevention programs as outgrowth of the Bogalusa Heart Study

- Health Ahead/ Heart Smart (K-6)
  Health Education for Elementary School Children
- Family Health Prevention Program
- Washington Parish (County) – wide
  Health Promotion for Children

made weekly, like weight, blood pressure, and lifestyle changes. It is especially important to identify children at high risk early from parents who have had heart attacks or cardiac events prior to age 60 years. Figure 11.6 illustrates results from the program.

This type of program should become an important component of clinical cardiology. Cardiologists should refer patients and spouse, and offspring for a rehabilitation and training program to help families understand lifestyle changes needed for prevention [21]. As described this is a high risk program to train individuals in families who have developed abnormal risk factors or after cardiac events. It is important to include family members since they share household and similar environmental exposure and have a related genetic background. Addressing families and offspring is another approach to beginning primordial prevention in children.

11.4.2 Health Promotion and Health Education for Children

With the realization of the tremendous prevalence of risk factors and the underlying cardiovascular diseases in our nation, (beginning in childhood), a pervasive public health strategy needs implementation. "Health Ahead/Heart Smart", K-6, education program, is one example [22, 23]. Although limited prevention and health messages are currently taught in schools—seat belts, helmets, and teeth brushing are taught—more in the line of heart disease, diabetes, tobacco, and alcohol use and poor life-

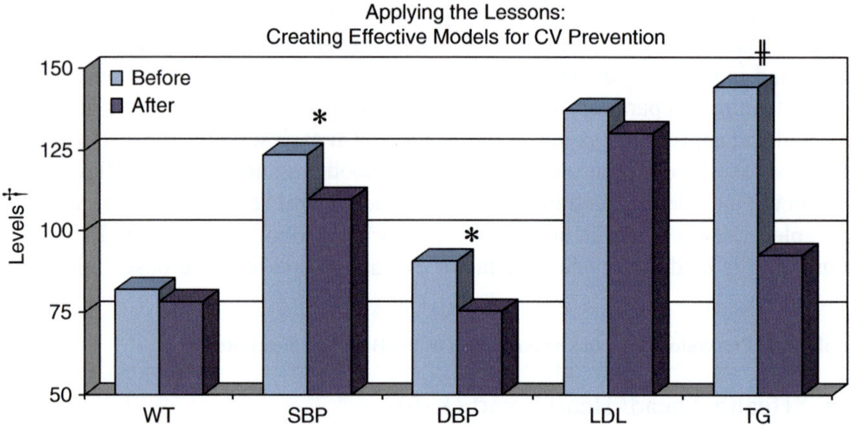

Fig. 11.6 The Heart Smart Family Health Promotion Program: comprehensive, interdisciplinary 12-wk intervention: education on CV risk, diet, exercise, behavior modification strategies. Significant improvement in CV risk factor levels in parents. Expanded to K-6, community centered intervention in New Orleans. [20]

* $p < 0.02$, ╫ $p < 0.05$

WT, weight; SBP, systolic blood pressure; DBP, diastolic blood pressure; LDL, low density lipoprotein cholesterol; TG, triglycerides

† Weight in kgs, blood pressure in mm Hg, Lipids in mg/dL

style behaviors are needed. We document consistent tobacco use in the third grade, at 8–9 years of age. Obviously poor lifestyles begin even earlier. Drop outs in education from finishing high school routinely are 35–45%. Teen age pregnancy, sexually transmitted diseases and violent behavior are common. Consequently, health education needs to be broad in scope.

The Health Ahead/Heart Smart format is a comprehensive and strongly behavioral-oriented program that addresses the entire school environment—the classroom, food service, physical education and programs for role models—teachers and parents. The program addresses health education broadly and is to be incorporated into the general curriculum. Beginning in kindergarten the program introduces self esteem, respect for their bodies, and encourages adopting healthy lifestyles. It stresses social and medical problems ie. obesity, tobacco use, dropouts, teen pregnancy, violent behavior, as well as nutrition and physical activity in a progressive scope and sequence. Importantly, lesson plans are strongly behavioral oriented, emphasizing healthy decision making. An afternoon health program for teachers is recommended and newsletters for parents and other activities in an effort to involve families in the education process. These are activities to encourage role models to be a part of education process. More detail is presented in Chap. 14.

11.4.3 A Parish (County) Wide Health Promotion for Children

An application of the broad health education program has been developed to cover an entire geographic area, a Parish or County [24]. The purpose is to involve the substructure of the area to implement health promotion in children. The Washington Parish (County) Health Prevention for Children incorporates the business, medical, and education aspects of the Parish as a model for other geographic areas. Table 11.2 shows various components of intervention in the community. Although the main intervention involves school children, the program attempts to address elements of the community to help with conducting prevention at a young age. The Parish program promotes an interaction with the medical community, and school nurse for health problems, like asthma-bronchitis. This overall program has been successful in helping to control obesity and improve physical activity in a Parish (County) wide effort. Involvement of the community ensures internalization of the program to provide interest and financial support to continue a program of prevention.

11.4.4 Role of Cardiologist

Knowing the late consequences and end stages of heart disease, Cardiologists can be instrumental in development of primordial prevention programs. Cardiologists have the knowledge and background to encourage and supervise health promotion programs for children and families.

Table 11.2 Intervention approaches

- School System

  Health Education, "Health Ahead/ Heart Smart" Program
- Medical

  Washington Parish Medical Society State Health Clinics School –based Clinics
- Social

  Parents and Volunteer Organizations
- Community

  Business
- News

  Media

Washington Parish (County)

11.5 Comment

It is important for Cardiologists to provide leadership to assist in efforts of prevention, particularly for involvement of families with known heart disease. Health education in the school system is a practical approach, to address heart disease as a broad public health initiative, but interest in the area of prevention beginning at a young age requires the effort of the community and the education system with support from medical personnel and business in the area.

References

1. Berenson GS et al (1980) Cardiovascular risk factors in children: the early natural history of atherosclerosis and essential hypertension. Oxford University Press, New York
2. Berenson GS et al (1986) Causation of cardiovascular risk factors in children: perspectives on cardiovascular risk in early life. Raven Press Books, New York
3. Hetzel BS, Berenson GS (1987) Cardiovascular risk factors in childhood: epidemiology and prevention. Elsevier, Amsterdam
4. Lauer RM, Burns TL, Daniels SR (2006) Pediatric prevention of atherosclerotic cardiovascular disease. Oxford University Press, New York

5. Viikari J, Akerblom HK, Räsänen L, Kalavainen M, Pietarinen O (1990) Cardiovascular risk in young Finns. Experiences from the Finnish multicentre study regarding the prevention of coronary heart disease. Acta Paediatr Scand Suppl 365:13–19
6. The top ten causes of death worldwide. The World Health Organization Statistical Update. Fact sheet #310. November 2008
7. Olshansky SJ, Passaro DJ, Hershow RC et al (2005) A potential decline in life expectancy in the United States in the 21st century. N Engl J Med 352:1138–1145
8. Lauer RM, Lee J, Clarke WR (1989) Predicting adult cholesterol levels from measurements in childhood and adolescence: the Muscatine Study. Bull N Y Acad Med 65:1127–1160
9. Franks PW, Hanson RL, Knowler WC et al (2007) Childhood predictors of young-onset type 2 diabetes. Diabetes 56:2964–2972
10. Unger RH, Scherer PE (2010) Gluttony, sloth and the metabolic syndrome: a roadmap to lipotoxicity. Trends Endocrinol Metab 21:345–352
11. Toprak A, Wang H, Chen W, Paul T, Ruan L, Srinivasan S, Berenson G (2009) Prehypertension and black–white contrasts in cardiovascular risk in young adults: Bogalusa Heart Study. J Hypertens 27:243–250
12. Nguyen QM, Srinivasan SR, Xu JH, Chen W, Berenson GS (2010) Fasting plasma glucose levels within the normoglycemic range in childhood as a predictor of prediabetes and type 2 diabetes in adulthood: the Bogalusa Heart Study. Arch Pediatr Adolesc Med 164:124–128
13. Cohn JN, Duprez DA, Grandits GA (2005) Arterial elasticity as part of a comprehensive assessment of cardiovascular risk and drug treatment (Use of the Rasmussen score). Hypertension 46:217–220
14. Stein JH, Korcarz CE, Hurst RT et al (2008) Use of carotid ultrasound to identify subclinical vascular disease and evaluate cardiovascular disease risk: a consensus statement from the American Society of Echocardiography Carotid Intima-Media Thickness Task Force. J Am Soc Echo 21:93–111
15. Urbina EM, Srinivasan SR, Kieltyka RL, Tang R, Bond MG, Chen W, Berenson GS (2004) Correlates of carotid artery stiffness in young adults: the Bogalusa Heart Study. Atherosclerosis 176:157–164
16. WHO Expert Committee (1990) Prevention in childhood and youth of adult cardiovascular diseases: Time for action. Geneva, Switzerland WHO: Tech Report Series 792
17. Rose G (1985) Individuals in sick populations. Int J Epidemiol 14:32–38
18. Wynder EL (1994) Principles of disease prevention from discovery to application. Soz Praventivmed 39:267–272
19. Berenson GS, Harsha DW, Johnson CC (1993) Teach families to be heart smart. Patient Care 6:134–135
20. The Bogalusa Heart Study. 20th Anniversary Symposium (1995) Am J Med Sci 310 (Suppl 1): S1–S138
21. Berenson GS, Srinivasan SR, Fernandez C, Chen W, Xu J (2010) Can adult cardiologist play a role in prevention of heart disease beginning in childhood? Methodist Debakey Cardiovasc J 6:4–9
22. Downey AM, Greenberg JS, Virgilio SJ, Berenson GS (1989) Health promotion model for "Heart Smart": the medical school, university, and community. Am J Health Promotion 13:31–46
23. Downey AM, Frank GC, Webber LS, Harsha DW, Virgilio SJ, Franklin FA, Berenson GS (1987) Implementation of "Heart Smart:" a cardiovascular school health promotion program. J Sch Health 57:98–104
24. Berenson GS (2010) Cardiovascular health promotion for children: a model for a Parish (County)-wide program (implementation and preliminary results). Prev Cardiol 13:23–28
25. Broyles S, Katzmarzyk PT, Srinivasan SR, Chen W, Bouchard C, Freedman DS, Berenson GS (2010) The pediatric obesity epidemic continues unabated in Bogalusa, Louisiana. Pediatrics 125:900–905
26. Srinivasan SR, Myers L, Berenson GS (1999) Temporal association between obesity and hyperinsulinemia in children, adolescents, and young adults: the Bogalusa Heart Study. Metabolism 48:928–934

10. [illegible] Srinivasan SR [illegible] Berenson GS [illegible] the Bogalusa Heart Study [illegible]
18. [illegible]

Chapter 12
Dietary Intake of Children over Two Decades in a Community and an Approach for Modification

Theresa A. Nicklas and Carol E. O'Neil

Abstract Dietary intake is a major environmental factor influencing health and disease. Dietary studies of intake and eating patterns are a cornerstone of cardiovascular (CV) research. Such studies have been conducted in the Bogalusa Heart Study from its origin and have helped understand childhood nutrition and its influence on CV risk factors and adult CV diseases.

Adverse high levels of calories, sodium, saturated fat, cholesterol, and refined sugar have been noted and eating patterns associated with overweight and the metabolic syndrome are described. Despite careful methodology, documenting the amount of excess daily energy intake over time that can result in overweight is limited. Observations of secular trends showed a significant negative trend in energy intake relative to body weight, a decrease of cholesterol intake, and a positive trend of total carbohydrate and starch intake. Beverage consumption has increased and has contributed to increased total energy intake, while fruit/juice/vegetable intake was higher in subjects with lower risk factors.

An effort to improve intake of fruit and vegetable was conducted as an intensive media campaign for high school students and their parents. The "Gimme 5 Nutrition Concept" effectively helped to develop positive attitudes and improvement of dietary habits of high school students.

Keywords Nutrition • Diet • Calories • Eating patterns • Dietary changes over time • Secular trends • Diet education

12.1 Introduction

Many recognize the Greek physician Hippocrates as the father of epidemiology [1] since he was the first person known to have examined the relationships between the occurrence of disease and environmental influences. Major environmental influ-

T. A. Nicklas (✉)
Department of Pediatrics, Baylor College of Medicine, Houston, TX, USA
e-mail: tnicklas@bcm.edu

G. S. Berenson (ed.), *Evolution of Cardio-Metabolic Risk from Birth to Middle Age*,
DOI 10.1007/978-94-007-1451-9_12, © Springer Science+Business Media B.V. 2011

ences on health and disease are dietary intake and eating patterns. Dietary studies are a cornerstone of public health research, since they help to identify dietary risk factors for disease and determine dietary treatment and prevention approaches for promoting optimal health.

For more than two decades the dietary studies conducted in the Bogalusa Heart Study (BHS) have made major contributions to the scientific literature by broadening our understanding of child nutrition and its influence on cardiovascular risk factors and obesity. As the original cohort of the BHS has aged, more recent studies have looked at eating patterns of young adults as they relate to metabolic syndrome, a major risk factor for cardiovascular disease. The methodology used in the dietary studies was reported in the first book published on the BHS [2]. In the second book [3] the focus was on studies looking at dietary intake as a determinant of cardiovascular risk factors. The current chapter in this third book, summarizes the significant milestones that the dietary studies of the BHS have made to child nutrition. These have included secular trends in dietary intake and food consumption patterns from 1973 to 1994; changes in food consumption patterns from childhood to young adulthood; and eating patterns associated with overweight. Follow-up studies of the original cohort have also looked at food patterns associated with weight and metabolic syndrome in young adults. Data from the dietary studies have been instrumental for generating hypotheses; designing intervention programs; developing public health policy; and, contributing to the science base for developing the Dietary Guidelines for Americans and the American Dietetic Association position paper on Dietary Guidance for Healthy Children 2–11 years of age (y) [4].

The quality of data from epidemiologic studies depends on the training of personnel and adherence to rigid protocols. It also depends on the validity and reliability of the test instruments used as well as on the responses of the subjects. For example, in the BHS the 24 h diet recall method had to be adapted for use in children [5, 6]. To improve the reliability and validity of the 24 h diet recall, quality controls included the use of standardized protocol that specified the exact techniques for interviewing, recording, and calculating results; standardized graduated food models to quantify foods and beverages most commonly forgotten; school lunch assessment [7] to identify all school lunch recipes, preparation methods, and average portion sizes of menu items reflected in each 24 h diet recall; follow-up telephone calls to parents to obtain information on brand names, recipes, and preparation methods of meals served at home; products researched in the field to obtain updated information on ingredients and preparation, and their weights (primarily snack foods and fast foods). All interviewers participated in rigorous training sessions and pilot studies before the field surveys to minimize interviewers effects. One 24 h diet recall was collected on each study participant, and duplicate recalls were collected from a 10% random subsample to assess interviewer variability [8]. Dietary data were collected on adolescents and young adults using validated food frequency questionnaire [9, 10]. These data allowed us to examine the relation between food patterns and weight and metabolic syndrome.

12.2 Trends in Nutrient Intake of 10-Year-Old Children over Two Decades (1973–1994) [11]

Although there has been a decline of coronary heart disease (CHD) over the past two decades, CHD remains the major cause of death in the US [12]. Prudent dietary habits can reduce the risk of CHD [13–18]. Population-based studies have documented changes in dietary intake and risk factors for cardiovascular disease over the past three decades [19–26]. The majority of these studies have focused on adults and only limited data have documented secular dietary trends in children. As part of the BHS, dietary intakes of children 10 years of age were examined in seven cross-sectional surveys which allowed us to determine secular trends in nutrient intake over two decades (1973–1994). These multiple cross-sectional studies provided the opportunity to look at longer term changes in the diet of children.

Intake of total energy, macronutrient, and cholesterol for each of the seven survey years from 1973 to 1994 is presented in Table 12.1. Total energy intake remained virtually the same over these years, ranging from 2,054 to 2,224 kilocalories (kcal). A significant negative trend was observed in energy intake relative to body weight. No significant trend was seen in total protein intake. A significant positive trend was noted in total carbohydrate and starch intake. Although total sugar (simple carbohydrates) intake did not change over the study period, a significant decrease in sucrose intake and a significant increase in fructose intake occurred.

A significant negative trend was observed with respect to intake of total fat, saturated fatty acids (SFA), and monounsaturated fatty acids (MUFA). The trend was opposite for polyunsaturated fatty acid (PUFA) intake. Cholesterol intake decreased significantly over the years, ranging from 324 (1973–1974) to 246 mg (1992–1994).

The percent contribution of the macronutrients to energy intake is presented in Table 12.2. Percent of energy from protein and carbohydrate increased significantly over the years. In contrast, the percent of energy from fat decreased significantly. This decrease was reflected in percent of energy from SFA and MUFA. Percent energy from PUFA showed a significant positive trend.

A visual comparison of dietary intakes of children 10 years of age in the BHS with children 6–11 years of age who participated in the 2007–2008 National Health and Nutrition Examination Survey (NHANES) suggested that intakes of selected nutrients have continued to change since 1994. The most obvious trend was the decrease in the absolute mean intake of total fat, SFA, and cholesterol. The mean intake of dietary cholesterol decreased from 246 to 234 mg and 195 mg for males and females, respectively. Relative to energy intake, the percent energy from total fat decreased from 36 (1992–1994) to 33% (2007–2008), with a concomitant increase in total carbohydrate intake from 51 to 55%. The decrease in total fat was most notable for intakes of palmitic and stearic fatty acids (Table 12.3). These trends in dietary intakes of children need to be confirmed in future analyses of the NHANES surveys from 1975 [27] to the present.

Table 12.1 Intake of selected nutrients of children 10 years of age in the Bogalusa Heart Study (1973–1994) (4), by survey year and compared with children 6–11 years of age participating in the National Health and Nutrition Examination Survey (NHANES) 2007–2008. [22]

Nutrient (g)	Survey years							P for linear trend	NHANES (6–11 years of age) 2007–2008	
	1973–1974 (N=185)	1976–1977 (N=158)	1978–1979 (N=224)	1981–1982 (N=304)	1984–1985 (N=284)	1987–1988 (N=284)	1992–1994 (N=216)		Males (n=550)	Females (n=571)
Energy (kcal)	2141[a]	2316	2145	2054	2145	2224	2116	NS	2042	1824
Protein	69.1	79.8	72.1	67.4	72.3	76.9	74.2	NS	70.8	61.9
Carbohydrate	262	281	253	256	272	285	270	0.029	276	245
Starch	88.4	94.6	88.2	85.9	86.5	91.1	74.6	0.047	–	–
Total sugars	144	155	134	131	152	160	141	NS	141	123
Total fat	93.1	100.5	96.1	87.3	88.0	88.6	84.6	0.003	75.2	68.6
Saturated	38.2	39.7	36.4	32.6	32.1	32.2	29.7	0.0001	26.9	24.0
Monounsaturated	–	–[b]	35.4	31.4	30.4	30.7	28.7	0.0002	27.6	24.8
Polyunsaturated	11.9	15.1	17.4	15.9	15.4	15.0	16.2	0.003	14.1	14.0
Cholesterol (mg)	324	322	317	266	240	285	246	0.0001	234	195

Bogalusa Heart Study data are adjusted for race and gender

[a] Mean

[b] Monounsaturated fatty acids were not available at the time of these survey

Table 12.2 Macronutrient composition of diet of 10-year-old children in the Bogalusa Heart Study (1973–1994) (4), by survey year and compared with children 6–11 years of age participating in the National Health and Examination Survey (NHANES) 2007–2008. [22]

Nutrients (% energy)	Survey years							P value for trend	NHANES (6–11 years of age) 2007–2008	
	1973–1974 (N=185)	1976–1977 (N=158)	1978–1979 (N=224)	1981–1982 (N=304)	1984–1985 (N=284)	1987–1988 (N=284)	1992–1994 (N=216)		Males	Females
Carbohydrate	49.4	49.6	49.2	50.9	51.4	51.5	51.3	0.002	55	54
Fat	38.4	38.3	38.5	37.4	36.3	35.6	35.8	0.0001	33	33
Saturated	15.9	15.2	14.7	14.0	13.2	13.0	12.5	0.0001	12	12
Polyunsaturated	4.87	5.78	6.98	6.85	6.41	6.13	7.15	0.0001	6	7
Monounsaturated	–	–	14.1	13.3	12.5	12.3	11.9	0.0001	12	12
Protein	13.0	13.5	13.3	13.1	13.5	13.9	13.9	0.001	14	14

Table 12.3 Fatty acid intake of children 10 years of age in the Bogalusa Heart Study (1973–1994) (4), by survey year and compared with children 6–11 years of age participating in the National Health and Examination Survey (NHANES) 2007–2008. [22]

Fatty acid[b] (g)	Survey years							P value for trend[a]	NHANES (6–11 years of age) 2007–2008	
	1973–1974 (N=185)	1976–1977 (N=158)	1978–1979 (N=224)	1981–1982 (N=304)	1984–1985 (N=284)	1987–1988 (N=284)	1992–1994 (N=216)		Males	Females
Myristic	2.8	3.3	3.4	3.0	3.0	2.9	2.6	0.0008	2.5	2.2
Palmitic	20.3	20.7	19.7	17.2	17.1	17.2	15.5	0.0001	14.2	12.6
Stearic	10.4	11.2	8.5	7.7	8.3	8.2	7.2	0.0001	6.8	6.1
Oleic	33.9	36.4	33.1	29.9	28.0	28.3	25.7	0.0001	25.7	23.3
Linoleic	11.4	13.6	15.6	13.9	13.2	12.9	14.0	0.0001	12.6	12.6
Linolenic	0.37	1.30	1.48	1.40	1.28	1.30	1.33	NS	1.12	1.08
Arachidonic	0.14	0.18	0.24	0.24	0.19	0.18	0.18	NS	0.10	0.08

[a] Adjusted for race and gender
[b] Mean ± SD

12.3 Children's Meal Patterns Have Changed over a 21-Year Period [28]

An important factor influencing the general health and well-being of children is the pattern of meal consumption. Meal patterns, a term that encompasses whether a meal was eaten, where it was eaten, its time span, and the number of eating episodes per day, have been implicated in obesity [29–34], cholesterol lipoprotein levels [35–42], glucose metabolism [37–40, 43], plasma hormones [44], energy density to energy intake [45], and nutrient use [46]. The BHS dietary studies provided the opportunity to examine changes in meal patterns of children over a 21-year period.

The School Breakfast Program was authorized in 1966; however, the program was available in Bogalusa, LA elementary schools until 1979. From 1973 to 1979, the percentage of children consuming a home breakfast decreased significantly from 86 (1973–1974) to 68% (1978–1979) ($p<0.001$); both racial and gender groups showed this decrease. With this trend, there was an increase in the percentage of children skipping breakfast—from 8 (1973–1974) to 30% (1978–1979) ($p<0.001$). In 1981, the National School Breakfast Program was introduced in the Bogalusa school system. At that time, a substantial number of children began eating school breakfast.

From 1973–1974 to 1993–1994, the percentage of children consuming a school lunch decreased significantly from 90 to 78% ($p<0.001$). In contrast, the percentage of children bringing lunch from home increased from 1 to 11% ($p<0.001$), particularly among Blacks ($p<0.05$).

From 1973–1974 to 1993–1994, the percentage of children consuming dinner at home decreased significantly ($p<0.01$) from 89 to 76%. In contrast, during that time period, the percentage of children consuming dinner outside the home increased significantly from 5 to 19% ($p<0.01$). The percentage of children consuming any meals at restaurants significantly increased from 0.3 (1973–1974) to 5.4% (1993–1994) ($p<0.0001$). The mean number of meals consumed per day significantly ($p<0.001$) decreased from 3.01 to 2.81.

There was no significant trend in the mean amount of breakfast consumed from 1973 to 1994. The mean amount of lunch and dinner consumed significantly increased ($p<0.0001$); and the amount of snacks consumed significantly decreased ($p<0.0001$). These changes resulted in a significant change in total amount of food/beverage consumed in a 24 h period ($p<0.01$). Breakfast, on average, represented 15% of the total amount of food consumed. Lunch represented 23 to 27% and dinner represented 27 to 33% of the total amount of food consumed, depending on the study year. Snacks represented 22 to 34% of the total amount consumed.

The percentage of children consuming at least one snack in a 24 h period significantly decreased ($p<0.0001$) from 1973 to 1994. Of those children consuming snacks, there was a significant increase in the percentage consuming 1–2 snacks ($p<0.0001$) compared to a significant decrease in the percentage consuming 4 ($p<0.05$) or ≥ 5 snacks ($p<0.0001$). In 1973–1974, 24.7% of children consumed one to two snacks and 30.2% consumed five or more snacks compared to 52.2% and 7.9%, respectively, in 1993–1994.

The mean number of total eating episodes decreased from 6.6 (range 2–11) (1973–1974) to 5.2 (1993–1994) ($p<0.0001$). When assessing the mean number of eating episodes that occurred before 11 a.m., between 11 a.m and 4 p.m., and after 4 p.m., there was a significant linear decrease ($p<0.0001$) from 1973–1974 to 1993–1994, in mean number of eating episodes in all three time periods for both genders and ethnicities. For all surveys combined, males had significantly ($p<0.001$) fewer eating episodes (mean=2.4) between 11 a.m. and 4 p.m. than females (mean=2.6). After 4 p.m., Whites (mean=2.3) had significantly ($p<0.05$) more mean eating episodes than Blacks (mean=2.1).

The mean mealtime span significantly ($p<0.0001$) decreased from 12.4 (1973–1974) to 11.5 h (1993–1994). The shorter mealtime span resulted from a significantly ($p<0.001$) later time when the first meal/beverage was consumed and a significantly ($p<0.05$) earlier time when the last meal/beverage of the day was consumed. Across all surveys, males ($p<0.05$) and Blacks ($p<0.001$) had longer mealtime spans than females or Whites. No significant associations were found between meal patterns and overweight status.

12.4 Children's Food Consumption Patterns Have Changed over Two Decades [47]

The nutrient intake of children has changed positively over the past two decades (1973–1994) [11, 19, 21, 24, 25, 48, 49]. The percentage of energy intake from protein and carbohydrate increased and the percentage of energy from total fat decreased. The logical next step was to examine secular trends in food consumption patterns of children over the same time period. Identifying these trends in children's food consumption patterns provides the focus for the design of population-based behavioral strategies for the promotion of health and prevention of chronic disease.

The percentage of children 10 years of age consuming fats/oils ($p<0.0001$), vegetables ($p<0.01$), desserts ($p<0.0001$), candy ($p<0.0001$), condiments ($p<0.001$), eggs ($p<0.001$), milk ($p<0.0001$), sweetened beverages ($p<0.05$), and beef ($p<0.01$) significantly decreased from 1973 to 1994. In contrast, the percentage of children consuming fruit/fruit juices ($p<0.01$), mixed meats ($p<0.01$), poultry ($p<0.0001$), and cheese ($p<0.0001$) significantly increased from 1973 to 1994.

The mean amount consumed at school and at restaurants significantly increased ($p<0.0001$) between 1973 and 1994 and decreased ($p<0.0001$) at other places, including the home. Table 12.4 shows mean food group consumption from 1973 to 1994 examined in absolute gram amounts. The mean consumption of cheese ($p<0.0001$), sweetened beverages ($p<0.01$), fruit and fruit juices ($p<0.01$), mixed meats ($p<0.0001$), poultry ($p<0.0001$), salty snacks ($p<0.001$), and condiments ($p<0.0001$) significantly increased from 1973 to 1994. In contrast, the mean consumption of milk ($p<0.01$), breads/grains ($p<0.05$), pork ($p<0.01$), eggs ($p<0.001$), desserts ($p<0.0001$), candy ($p<0.0001$), and fats/oils ($p<0.05$) decreased.

Table 12.4 Mean (gram) food group consumption from 1973–1974 to 1993–1994

Food group	Study year (mean)						
	1973–1974	1976–1977	1978–1979	1981–1982	1984–1985	1987–1988	1993–1994
Fats and oils*	29.8	24.2	24.9	18.6	19.5	22.1	19.9
Fruit and fruit juices**	104.1	117.8	119.2	151.9	116.2	154.5	149.4
Vegetable	146.5	159.6	134.9	172.4	177.6	182.7	134.1
Breads and grains*	193.3	196.5	197.4	174.3	183.7	188.8	170.2
Mixed meats†	32.6	42.3	51.3	57.5	71.1	89.1	55.1
Deserts†	89.1	86.8	84.1	64.1	59.8	57.2	56.0
Candy†	45.0	54.2	41.3	34.1	42.3	34.0	34.8
Poultry†	23.6	22.1	20.5	25.1	26.1	37.5	60.8
Salty snacks***	8.4	9.2	12.9	12.3	12.6	10.7	17.3
Seafood	6.5	7.8	10.8	12.2	9.1	5.2	11.8
Condiments†	5.4	4.3	3.6	6.5	9.6	6.9	12.7
Egg***	18.4	13.6	12.6	11.2	8.1	7.7	8.8
Milk**	426.2	494.4	385.7	408.5	372.8	396.4	362.4
Pork**	25.3	26.9	29.9	25.8	17.8	18.0	15.0
Cheese†	14.4	13.9	17.0	15.4	27.4	28.7	32.7
Beef	46.4	58.5	55.2	38.0	40.7	35.1	67.7
Sweetened beverages**	369.9	382.1	375.7	412.7	371.4	372.5	448.2

SD Standard deviation

P for linear trend: *$p<0.05$; **$p<0.01$; ***$p<0.001$; † $p<0.0001$

12.5 Changes in Food Group Consumption Patterns from Childhood to Young Adulthood [50]

Learning how patterns of food consumption develop during childhood is important to understanding the pathogenesis of several chronic diseases later in life [51]. An important linkage in this chain is whether eating patterns track from childhood to adulthood. Examining changes in eating patterns from childhood to young adulthood may provide new insight into the linkage between food patterns during this age transition.

The percentage of children and young adults consuming food groups is presented in Table 12.5. At 10 years of age, the percentage of children consuming vegetables, breads/grains, poultry ($p<0.05$), mixed meats, desserts, fruit/fruit juice, candy, and milk ($p<0.001$) was significantly higher than the percentage consuming those food groups in young adulthood. There was a higher percentage of young adults consuming cheese and seafood than in childhood. It is unclear how many of these changes in food patterns reflect secular changes or changes during this age transition. There were no gender or ethnic differences in the percentage consuming food groups at both age periods.

There were differences in mean food group consumption from childhood to young adulthood. At 10 years of age, more fruits/fruit juices, mixed meats, desserts, candy, and milk were consumed than in young adulthood. In contrast, more sweetened beverages, poultry, cheese, seafood, salty snacks, and beef were consumed in young

Table 12.5 Percentage consuming food groups in childhood and young adulthood

	Childhood n (%)	Young adulthood n (%)	p-value
Fats/oils	200 (81.3)	190 (77.2)	NS
Fruit/fruit juices	160 (65)	74 (30.1)	p<0.001
Vegetables	218 (88.6)	200 (81.3)	p<0.05
Breads/grains	242 (98.4)	233 (94.7)	p<0.05
Mixed meats	104 (42.3)	52 (21.1)	p<0.001
Desserts	183 (74.4)	106 (43.1)	p<0.001
Candy	229 (93.1)	143 (58.1)	p<0.001
Poultry	73 (29.7)	96 (39)	p<0.05
Snacks	88 (35.8)	85 (34.6)	NS
Seafood	18 (7.3)	47 (19.1)	p<0.001
Condiments	146 (59.3)	152 (61.8)	NS
Egg	55 (22.4)	40 (16.3)	NS
Milk	224 (91.1)	111 (45.1)	p<0.001
Pork	91 (37)	100 (40.7)	NS
Cheese	46 (18.7)	100 (40.7)	p<0.001
Beef	128 (52)	132 (53.7)	NS
Sweetened beverages	240 (97.6)	240 (97.6)	NS

NS Not Significant at $p<0.05$
NA Not Applicable (% did not change)
Comparing the change in percentage consuming food groups from childhood to young adulthood

Table 12.6 Distribution of Nutrient-Dense Scores from childhood to young adulthood

High Nutrient-Dense Scores[a]	Childhood n (%)	Young adulthood n (%)
One	0 (0)	3 (1.2)
Two	2 (0.8)	24 (9.8)
Three	22 (8.9)	72 (29.3)
Four	98 (39.8)	101 (41.1)
Five	124 (50.4)	46 (18.7)
Total gram amount[b]	*1,027*	*920*

[a] Nutrient-Dense Scores reflected foods consumed at least once from the meats, dairy, breads/grains, fruit/fruit juices and vegetables food groups

[b] Total gram amount reflects consumption of the High-Nutrient-Dense foods

adulthood than in childhood. No association was found between food group consumption patterns and Body Mass Index (BMI) at any time points during the study.

From childhood to young adulthood, bread/grain consumption decreased for both Whites and Blacks, with a greater decrease among Blacks. Candy consumption decreased for both Whites and Blacks, with a greater decrease among Whites. Condiment consumption increased for both Whites and Blacks, with a greater increase among Blacks.

At 10 years of age, 50% of the children had a high nutrient-dense (HND) score of 5 and only 19% received a HND score of 5 in young adulthood (Table 12.6) suggesting that the nutrient density of the diet decreased during this age transition. Only 12% of the sample consumed a food from each of the 5 food groups (*e.g.*, meats, dairy, breads/grains, fruit/fruit juices, and vegetables) at both baseline (10 years of age) and follow-up (young adulthood). Of those children 10 years of age who consumed foods from 4 of the 5 HND food groups, most fell short of consuming fruit/fruit juices (67%) or vegetables (17%). At young adulthood, most fell short of consuming fruit/fruit juices (83%) or dairy (9%). In childhood, the mean amount of HND foods was 1,027 g but was 920 g in young adulthood.

12.6 Eating Patterns and Overweight Status in Children [52] and Young Adults [53]

Analyses were also conducted to see if there was a relationship between the frequency and mean gram consumption of food groups from childhood to young adulthood. The mean frequency of food groups consumed decreased significantly from childhood to young adulthood regardless of gender and ethnicity. For an entire 24 h period, the mean amount of total food consumed in childhood was 1,474 g and in adulthood was 1,803 g.

Obesity among children has increased dramatically over the past three decades [54–56]. Using data from the NHANES 2007–2008, it has been shown that 31.7% of children are overweight BMI ≥ 85th percentile) and 16.9% are obese (BMI ≥ 95th percentile). These figures are nearly triple what they were in 1980.

The dietary influences on obesity are complex and poorly understood [57]. While individual nutrients have been implicated in obesity [58–62], several studies have looked at the association between eating patterns and weight. Studies have shown an association between restaurant food consumption [63–66], soft drink consumption [67, 68], increased portion sizes [69], meal patterns and meal frequency [29–32], diet quality [70], and food diversity [65] and BMI. However, most of these studies were conducted with adults and the findings have yet to be replicated in children.

Using BHS data, we examined the association between eating patterns and overweight status in children (Table 12.7). Total gram amount of food/beverage consumed, particularly from snacks, and total gram consumption of low quality foods were positively associated with overweight status. Consumption of sweetened beverages ($p<0.001$) were positively associated with overweight status in Whites only. In Black females, consumption of fruit/fruit juices was negatively associated with overweight status. Despite these significant associations, the percentage of variance explained by the model was very small. In the food group consumption model, only 5% of the variance was explained, of which sweetened beverages alone explained 1%. Similarly, the percent of variance explained by total gram amount of low quality foods consumed; total gram amount of foods/beverages consumed, particularly from lunch and dinner, and number of snacking episodes was equally low.

Eating patterns (*e.g.* food consumption and meal patterns) associated with being overweight among young adults were also identified (Table 12.8). The percent gram consumption of fruit/fruit juices was negatively associated with being overweight ($p<0.05$) while the percent gram consumption of diet beverages was positively associated with being overweight ($p<0.05$) or obese ($p<0.05$). Correspondingly, normal weight individuals, when compared with overweight individuals, tended to have a higher percentage gram consumption of fruit and 100% fruit juices ($p<0.05$), and less percent gram consumption of diet beverages than obese individuals ($p<0.05$). Obese participants consumed more food/beverage than normal weight participants ($p<0.05$), because of the increased consumption of diet beverages.

Ironically, the positive association between beverage consumption and overweight resulted from the consumption of diet beverages and non-sweetened beverages. This finding has also been reported in another study using national data [71]. The positive association between sweetened beverages and overweight in children and adults continues to be debated [72]. Studies supporting an association found that an increased consumption of sweetened beverages was associated with increased total energy intake [73], particularly among overweight children and adolescents.

It is important to note that individual eating patterns explained only 1–2% of the variance in BMI. However, since the eating patterns studied were not mutually exclusive, it was impossible to determine the cumulative effect of the significant eating patterns on overweight status. It can be hypothesized, however, that the association between eating patterns and overweight status is not a result of a single eating pattern, but of the combination of eating patterns that are interrelated and cumulative in their effect on overweight status. Moreover, these eating patterns associated with overweight may also vary by gender and ethnicity, further complicating the picture [52].

Table 12.7 The association between eating-pattern variables and overweight status by ethnicity–gender groups

Eating pattern	White males OR (95% CI)	White females OR (95% CI)	Black males OR (95% CI)	Black females OR (95% CI)
Food groups consumption I[a,b]	R2=0.08	R2=0.10	R2=0.17	R2=0.13
Fats/oils	0.97 (0.85–1.10)	1.00 (0.83–1.19)	0.93 (0.71–1.22)	1.06 (0.86–1.32)
Fruit/fruit juices	1.03 (0.88–1.20)	1.10 (0.92–1.31)	0.97 (0.69–1.41)	0.55 (0.38–0.79)*
Vegetables	0.98 (0.77–1.24)	1.09 (0.87–1.36)	1.05 (0.74–1.49)	0.75 (0.51–1.09)
Breads/grains	1.20 (0.86–1.67)	0.90 (0.62–1.30)	0.62 (0.33–1.16)	1.03 (0.60–1.79)
Mixed meats	1.12 (0.95–1.31)	0.93 (0.78–1.12)	1.06 (0.82–1.37)	0.97 (0.78–1.19)
Desserts	0.89 (0.73–1.09)	1.08 (0.86–1.35)	0.89 (0.65–1.22)	0.89 (0.66–1.21)
Candy	0.94 (0.76–1.18)	0.78 (0.60–1.01)	0.79 (0.51–1.23)	1.00 (0.73–1.35)
Sweetened beverages	1.68 (1.21–2.33)*	1.53 (1.05–2.22)*	1.02 (0.72–1.46)	0.92 (0.65–1.30)
Poultry	0.99 (0.89–1.09)	1.04 (0.94–1.16)	0.97 (0.76–1.23)	0.99 (0.84–1.16)
Salty snacks	0.98 (0.88–1.09)	0.92 (0.80–1.05)	1.15 (0.94–1.42)	0.84 (0.66–1.06)
Seafood	0.97 (0.92–1.02)	1.07 (1.01–1.13)*	0.73 (0.48–1.11)	1.03 (0.95–1.11)
Condiments	1.02 (0.93–1.12)	0.99 (0.90–1.08)	1.02 (0.86–1.22)	0.89 (0.74–1.07)
Eggs	0.97 (0.88–1.06)	1.07 (0.99–1.17)	0.99 (0.85–1.14)	0.91 (0.79–1.05)
Milk	0.96 (0.64–1.46)	1.19 (0.82–1.73)	1.25 (0.68–2.30)	0.93 (0.55–1.59)
Pork	1.04 (0.94–1.14)	1.00 (0.89–1.12)	1.16 (0.99–1.35)	0.98 (0.84–1.13)
Cheese	1.02 (0.95–1.11)	1.04 (0.97–1.13)	0.95 (0.82–1.09)	0.90 (0.77–1.06)
Beef	1.08 (0.92–1.25)	1.06 (0.91–1.24)	1.11 (0.89–1.37)	1.02 (0.88–1.18)
Gram from high- and low-quality foods	R2=0.05	R2=0.07	R2=0.09	R2=0.06
High-quality foods[c]	1.44 (0.70–2.95)	1.89 (0.90–3.96)	1.06 (0.34–3.33)	0.34 (0.12–1.04)
Low-quality foods[d]	1.69 (1.12–2.53)*	1.60 (1.01–2.53)*	1.00 (0.62–1.63)	0.89 (0.56–1.42)
Gram amount	R2=0.05	R2=0.07	R2=0.09	R2=0.06
Total	3.17 (1.20–8.41)* R2=0.05	2.97 (1.03–8.52)* R2=0.07	1.03 (0.25–4.30) R2=0.11	0.33 (0.09–1.22) R2=0.07
From breakfast	1.14 (0.81–1.60)	1.18 (0.83–1.71)	0.67 (0.40–1.10)	0.62 (0.42–0.93)*
From lunch	1.20 (0.75–1.93)	1.38 (0.86–2.23)	1.03 (0.48–2.23)	0.91 (0.47–1.77)
From dinner	1.82 (1.13–2.91)*	1.24 (0.72–2.14)	0.81 (0.43–1.54)	0.95 (0.54–1.67)
From snacks	1.39 (1.00–1.95)	1.43 (0.96–2.14)	1.34 (0.81–2.20)	0.68 (0.41–1.13)
Eating episode[e]	R2=0.04	R2=0.06	R2=0.09	R2=0.06
Total	0.90 (0.79–1.03) R2=0.04	1.07 (0.93–1.24) R2=0.06	1.04 (0.85–1.28) R2=01.4	0.91 (0.76–1.08) R2=0.07
No. of meals	0.97 (0.63–1.50)	1.23 (0.80–1.89)	0.70 (0.38–1.33)	0.56 (0.33–0.95)*
No. of snacks	0.90 (0.78–1.03)	1.06 (0.91–1.23)	1.07 (0.87–1.31)	0.93 (0.78–1.12)

Black CI confidence interval, *White FJV* fruit/fruit juices and vegetables, meats, mixed meats, poultry, seafood, eggs, pork, and beef, *OR* odds ratio, sweets, desserts, candy, and sweetened beverages, dairy, milk and cheese

**p*_0.05; **p*_0.01

[a] Food group consumption I: individual food group consumption as eating pattern variables

[b] Odds ratio: risk of being overweight if increasing mean gram consumption

[c] High-quality foods: fruit/fruit juices, vegetables, breads/grains, meats, dairy

[d] Low-quality foods: fats/oils, sweets, salty snacks

[e] Odds ratio: risk of being overweight if having one more eating episode

Table 12.8 Association between eating patterns and weight status

	Odds ratio			Mean difference			
	Overweight vs normal weight	R2	Obese vs normal weight	R2	Normal weight	Overweight	Obese
Food groups I (gram percent)							
Fat	0.95 (0.84, 1.07)		0.96 (0.84, 1.10)		1.41 (0.12)	1.24 (0.18)	1.27 (0.21)
Fruit/fruit juices	0.96 (0.93, 0.99)	0.02	0.98 (0.96, 1.01)		6.36 (0.61)^{A}	3.60 (0.87)^{B}	4.80 (1.05)^{AB}
Vegetables	0.99 (0.97, 1.02)		1.01 (0.98, 1.03)		7.99 (0.58)	7.21 (0.83)	9.05 (1.00)
Breads/grains	1.00 (0.98, 1.03)		0.99 (0.95, 1.02)		10.21 (0.50)	10.36 (0.72)	9.64 (0.86)
Mixed meats	1.05 (0.99, 1.10)		1.04 (0.98, 1.11)		0.99 (0.25)	1.74 (0.36)	1.42 (0.43)
Dessert	0.99 (0.94, 1.04)		1.00 (0.95, 1.06)		2.26 (0.27)	2.17 (0.39)	2.30 (0.47)
Candy	0.99 (0.90, 1.08)		0.90 (0.78, 1.04)		1.33 (0.14)	1.25 (0.20)	0.85 (0.24)
Non-alcohol beverage	1.01 (0.99, 1.02)		1.01 (1.002, 1.03)	0.02	39.75 (1.37)^{A}	43.42 (1.97)^{A,B}	46.57 (2.36)^{B}
Diet beverage	1.02 (1.002, 1.03)	0.007	1.02 (1.001, 1.03)	0.01	2.57 (0.91)^{A}	5.56 (1.32)^{A,B}	6.80 (1.58)^{B}
Sweetened beverage	1.00 (0.99, 1.01)		1.01 (0.99, 1.02)		37.25 (1.54)	37.94 (2.23)	39.84 (2.67)
Poultry	0.99 (0.96, 1.03)		1.01 (0.98, 1.04)		3.59 (0.46)	3.32 (0.66)	4.06 (0.79)
Snacks	0.95 (0.83, 1.08)		0.90 (0.76, 1.06)		0.86 (0.11)	0.77 (0.16)	0.56 (0.19)
Seafood	0.98 (0.94, 1.03)		0.98 (0.92, 1.03)		2.01 (0.34)	1.60 (0.50)	1.27 (0.59)
Condiments	1.09 (0.87, 1.38)		1.02 (0.77, 1.35)		0.50 (0.06)	0.57 (0.08)	0.52 (0.10)
Eggs	0.94 (0.83, 1.06)		0.93 (0.81, 1.07)		0.86 (0.13)	0.74 (0.18)	0.64 (0.22)
Milk	1.00 (0.98, 1.02)		1.00 (0.98, 1.03)		7.21 (0.75)	7.13 (1.09)	7.66 (1.30)
Pork	0.94 (0.87, 1.02)		0.93 (0.84, 1.03)		1.68 (0.19)	1.19 (0.28)	1.11 (0.33)
Cheese	1.00 (0.96, 1.04)		0.94 (0.87, 1.02)		2.11 (0.30)	2.04 (0.43)	1.17 (0.52)
Beef	1.03 (0.99, 1.07)		0.98 (0.93, 1.03)		3.43 (0.36)	4.51 (0.52)	3.09 (0.62)
Food groups II (gram percent)							
FJV	1.00 (0.99, 1.00)		1.00 (0.99, 1.001)		14.34 (0.84)^{A}	9.05 (1.20)^{B}	12.79 (1.44)^{AB}
Dairy	1.00 (0.98, 1.02)		1.00 (0.98, 1.02)		9.32 (0.80)	9.17 (1.16)	8.83 (1.38)
Meats	1.00 (0.98, 1.03)		0.99 (0.97, 1.02)		12.55 (0.64)	13.10 (0.93)	11.59 (1.11)

Table 12.8 (continued)

	Odds ratio			Mean difference			
	Overweight vs normal weight	R2	Obese vs normal weight	R2	Normal weight	Overweight	Obese
Sweets	1.00 (0.99, 1.01)		1.00 (0.99, 1.01)		40.84 (1.55)	41.36 (2.24)	42.99 (2.68)
Diet diversity score	0.89 (0.69, 1.14)		0.81 (0.62, 1.08)		3.66 (0.06)	3.57 (0.08)	3.53 (0.10)
Meal period							
Breakfast	0.99 (0.97, 1.01)		1.01 (0.99, 1.03)		14.04 (0.87)	12.40 (1.25)	16.18 (1.50)
Lunch	1.01 (0.99, 1.02)		0.99 (0.98, 1.00)		25.76 (1.17)	28.41 (1.70)	22.48 (2.03)
Snack	1.00 (0.99, 1.01)		1.00 (0.98, 1.01)		29.62 (1.37)	30.41 (1.97)	27.46 (2.36)
Dinner	0.99 (0.98, 1.01)		1.01 (0.99, 1.02)		30.58 (1.08)	28.78 (1.56)	33.87 (1.86)
Total gram amount	1.00 (0.99. 1.01)		1.00 (1.00, 1.001)		2121.16 (48.30)^{A}	2117.32 (69.77)^{AB}	2346.70 (83.51)^{B}
Eating episodes							
Number of meals	0.82 (0.60, 1.13)		0.91 (0.64, 1.30)		2.47 (0.04)	2.38 (0.06)	2.43 (0.08)
Number of snacks	0.97 (0.84, 1.12)		00.85 (0.70, 1.02)		2.37 (0.09)	2.30 (0.14)	2.01 (0.16)
Total eating episodes	0.93 (0.80, 1.08)		0.83 (0.69, 1.00)		4.84 (0.09)	4.68 (0.14)	4.44 (0.16)
Restaurant consumption	1.04 (0.99, 1.01)		1.01 (0.99, 1.02)		16.37 (1.59)	18.33 (2.31)	18.94 (2.75)

Model adjusted for age, calorie intake, ethnicity, gender, and ethnicity × gender. Data presented as the odds ratio (95% confidence interval) or as the least-square mean (standard error). Data with different superscript letters are statistically different

12.7 Comparison of Food Group Intake with Metabolic Syndrome in Young Adults [74]

Insulin resistance, central obesity, dyslipidemia, and hypertension are risk factors for coronary heart disease in adults, and the clustering of these risk factors is used to define metabolic syndrome [75]. In the US, approximately 36% of males and 32% of females meet the criteria for metabolic syndrome [76–81]. Not only do cardiovascular risk factors, including those associated with metabolic syndrome, and weight track into adulthood [77–80], but with the escalating pediatric obesity crisis, more children, adolescents, and young adults are being diagnosed with metabolic syndrome [81]. Most components of metabolic syndrome are related separately to lifestyle factors such as weight control [82], diet [82–87], and physical activity [88–90]. Most studies that examined the relationship between diet and metabolic syndrome risk factors have focused on nutrient-based, not food-based, analyses [91], assessed the relations with individual risk factors, not with risk factor clusters, and were conducted with middle-aged and older adults, not with a biracial population of young adults. The BHS dietary studies provided an opportunity to examine the relation between food group intakes and the number of metabolic syndrome risk factors in a well-studied biracial young adult population. Dietary intakes across the three metabolic syndrome risk factor groups are presented in Model 1 (Table 12.9). Model 2 was adjusted for age, sex, and ethnicity; Model 3 was further adjusted for total energy; BMI and level of physical activity were further adjusted in Model 4.

There were no significant differences in mean total daily energy intake across the three risk factor groups, even after adjusting for age, gender, ethnicity, BMI, and physical activity. Models 1, 2, and 3 also showed no difference in mean intake of low-fat dairy products. However, mean intake of low-fat dairy products was higher in subjects who had no risk factors than in subjects who had one to two risk factors (0.73 compared with 0.56 servings/day) after adjustment for age, sex, ethnicity, total energy, BMI, and physical activity.

No differences in mean intake of fruit/juice/vegetables (FJV) across the 3 risk factor groups were found in Models 1 or 2. However, mean intakes of FJV were higher in subjects who had no risk factors than in subjects who had one to two risk factors in Models 3 and 4 (3.29 compared with 3.01 servings/d in Model 3; 3.30 compared with 2.99 servings/d in model 4). Mean intakes of diet beverages were significantly higher in subjects who had no risk factors than in subjects who had one to two risk factors after adjustment for all covariates (0.76 compared with 0.62 servings/d).

12.8 Dietary Patterns Associated with Metabolic Syndrome in Young Adults [92]

Studies of dietary patterns (DP) and their association with disease have several benefits over the conventional approach, which has focused largely on the effects of single nutrients or individual foods [93, 94]. As dietary assessment is complex,

Table 12.9 Daily intakes of energy and of foods from various food groups according to number of metabolic syndrome risk factors among participants in the Bogalusa Heart Study

	No risk factors (n=468)	1–2 risk factors (n=571)	≥3 risk factors (n=142)	P
Total energy (kcal)				
Model 1	2115.02	2114.23	2132.01	0.96
Model 2	2101.28	2125.41	2132.36	0.81
Model 3	N/A	N/A	N/A	
Model 4	2087.40	2129.00	2143.12	0.64
Low-fat dairy products (servings)				
Model 1	0.67	0.56	0.60	0.24
Model 2	0.67	0.56	0.60	0.19
Model 3	0.68	0.56	0.59	0.14
Model 4	0.73[a]	0.56[b]	0.52[a,b]	0.03
High-fat dairy products (servings)				
Model 1	0.79	0.83	0.87	0.55
Model 2	0.77	0.84	0.88	0.28
Model 3	0.78	0.84	0.87	0.35
Model 4	0.73	0.85	0.95	0.04
Refined grains (servings)				
Model 1	1.68	1.61	1.59	0.53
Model 2	1.67	1.62	1.58	0.62
Model 3	1.68	1.61	1.57	0.31
Model 4	1.68	1.60	1.54	0.34
Whole grains (servings)				
Model 1	0.48	0.41	0.45	0.21
Model 2	0.48	0.42	0.44	0.25
Model 3	0.48	0.41	0.44	0.16
Model 4	0.48	0.41	0.43	0.20
Fruit, fruit juice, and vegetables (servings)				
Model 1	3.26±0.09	3.03	3.00	0.12
Model 2	3.27±0.09	3.03	2.96	0.07
Model 3	3.29±0.08[a]	3.01[b]	2.94[a,b]	0.01
Model 4	3.30±0.09[a]	2.99[b]	2.89[a,b]	0.02
Potatoes (servings)				
Model 1	0.32±0.01	0.31	0.34	0.41
Model 2	0.32±0.01	0.31	0.33	0.45
Model 3	0.32±0.01	0.31	0.33	0.37
Model 4	0.31±0.01	0.31	0.36	0.08
Meat (servings)				
Model 1	0.74	0.74	0.76	0.71
Model 2	0.73	0.75	0.74	0.83
Model 3	0.74	0.74	0.74	0.95
Model 4	0.75	0.74	0.74	0.98
Seafood (servings)				
Model 1	0.30	0.30	0.29	0.76
Model 2	0.30	0.30	0.29	0.65

Table 12.9 (continued)

	No risk factors (n=468)	1–2 risk factors (n=571)	≥3 risk factors (n=142)	P
Model 3	0.30	0.30	0.29	0.48
Model 4	0.30	0.30	0.30	0.84
Dishes with cheese (servings)				
Model 1	0.48	0.48	0.48	0.99
Model 2	0.47	0.49	0.49	0.76
Model 3	0.48	0.48	0.48	0.88
Model 4	0.49	0.48	0.46	0.59
Diet beverages (servings)				
Model 1	0.68	0.61	0.74	0.10
Model 2	0.70	0.62	0.68	0.14
Model 3	0.70	0.62	0.68	0.14
Model 4	0.76^{a}	0.62^{b}	$0.59^{a,b}$	0.01
Sweet snacks (servings)				
Model 1	1.29	1.38	1.32	0.49
Model 2	1.28	1.40	1.30	0.31
Model 3	1.29	1.39	1.28	0.31
Model 4	1.27	1.40	1.30	0.14

For total energy intake, the models were as follows: Model 1, unadjusted ANOVA; Model 2, adjustment for age, sex, and ethnicity; Model 3, not applicable (N/A) for total energy intake; Model 4, adjustment for age, sex, ethnicity, BMI, and physical activity. For intakes of foods from various food groups, the models were as follows: model 1, ANOVA; Model 2, adjustment for age, sex, and ethnicity; Model 3, adjustment for age, sex, ethnicity, and total energy intake; Model 4, adjustment for age, sex, ethnicity, total energy intake, BMI, and physical activity

Values in the same row with different superscript letters are significantly different, $P \leq 0.05$ (Bonferroni correction for post hoc multiple comparisons)

[1] All values are $x \pm SE$

and foods are typically consumed in combinations, the combined effect of nutrients and foods can be observed only when DP are examined [93, 94]. Results from DP analyses are more helpful in disseminating diet-related messages to consumers in that they may be more likely to adhere to DP rather than those related to single foods or nutrients [95]. DP have also been related to selected biomarkers of dietary exposure [93, 94] and have been reported to contribute in the development or retention of CHD and type 2 diabetes mellitus [96].

Two studies using BHS [74, 92] data have examined the relationship of diet with metabolic syndrome in young adults. Results from the first study [74] showed that lower fruit and vegetable consumption and higher sweetened beverage consumption were independently associated with one to two risk factors for metabolic syndrome [74]. A limitation of that analysis was that the relationship of metabolic syndrome and single food groups were explored, rather than DP. The second study aimed to identify various DP among young adults and to examine the association of these DP with the risk factors of metabolic syndrome.

Factor analysis retained two DP (Table 12.10): the 'Western Dietary Pattern' (WDP), consisting of refined grains, French fries, high-fat dairy products, dishes

Table 12.10 Identification of DP from factor loadings

Food items	Factor loadings[a]	
	WDP	PDP
1. Whole grains	–	0.46
2. Legumes	–	0.61
3. Cruciferous vegetables	–	0.70
4. Other vegetables	–	0.74
5. Green leafy vegetables	–	0.69
6. Dark-yellow vegetables	–	0.70
7. Tomatoes	–	0.58
8. Fruit	–	0.64
9. 100% fruit juices	–	0.43
10. Low-fat dairy products	–	0.36
11. Poultry	–	0.40
12. Clear soups	–	0.36
13. Low-fat salad dressings	–	0.49
14. Refined grains	0.43	–
15. French fries	0.53	–
16. High-fat dairy products	0.53	–
17. Dishes with cheese	0.58	–
18. Red meats	0.50	–
19. Processed meats	0.59	–
20. Eggs	0.39	–
21. Snacks	0.53	–
22. Sweets and desserts	0.54	–
23. Sweetened beverages	0.44	–
24. Condiments	0.40	–
Variability explained	19%	12%

DP dietary patterns, *YAQ* youth and adolescent food frequency questionnaire, *WDP* Western dietary pattern, *PDP* prudent dietary pattern

[a] Data (1–24) are factor loadings (correlation coefficients between the variables and factors) derived from principal component factor analysis

Absolute values of factor loadings, 0.30 are indicated by '–' for simplicity

with cheese, red meats, processed meats, eggs, snacks, sweets and desserts, sweetened beverages, and condiments and the 'Prudent Dietary Pattern' [PDP], consisting of whole grains, legumes, vegetables, tomatoes, fruit, 100% fruit juices, low-fat dairy products, poultry, clear soups, and low-fat salad dressings. The WDP and the PDP explained 19% and 12% of the dietary intake variance, respectively.

Covariate-adjusted associations between dietary patterns and components of metabolic syndrome are presented in Table 12.11. Using the covariate-adjusted model (excluding BMI), waist circumference, triceps skin fold, plasma insulin and the occurrence of metabolic syndrome were all inversely associated with the PDP. Insulin sensitivity was positively associated with the PDP. Serum triglycerides were negatively associated with both PDP and WDP. After adjusting for BMI in addition to other covariates, serum high density lipoprotein-cholesterol was inversely associated with the WDP. The overall prevalence of metabolic syndrome did not differ by the two DP.

Table 12.11 Covariate-adjusted associations between risk factors for MetS and DP in young adults (19–39 years)

Obesity measurements	Risk factors (dependent variables)							
	Model 1[a]				Model 2[b]			
	PDP (n 995)		WDP (n 995)		PDP (n 995)		WDP (n 995)	
	Std β	P value	Std β	P value	Std β	P value	Std β	P value
BMI[c]	−0.07	0.15	−0.05	0.42	–	–	–	–
Waist circumference	−0.10	0.02	−0.05	0.46	−0.04	0.009	0.00	1.00
Triceps skinfold	−0.11	0.01	−0.02	0.75	−0.06	0.17	0.02	0.62
Blood pressure								
Systolic	−0.05	0.29	−0.05	0.42	−0.03	0.51	−0.04	0.60
Diastolic	−0.01	0.80	0.00	0.99	0.01	0.86	0.17	0.80
Diabetes mellitus								
Plasma glucose	−0.03	0.52	−0.05	0.49	−0.01	0.83	−0.03	0.64
Plasma insulin	−0.10	0.03	−0.04	0.53	−0.07	0.09	−0.01	0.83
Insulin resistance: HOMA-IR	−0.07	0.11	−0.04	0.58	−0.04	0.33	−0.01	0.89
Insulin sensitivity: QUICKI	0.16	<0.0005	0.02	0.78	0.13	0.001	−0.01	0.82
Lipid profiles	−0.01	0.88	−0.08	0.22	0.00	0.94	−0.08	0.27
Total serum cholesterol								
Serum LDL-C	−0.01	0.91	−0.02	0.77	0.01	0.92	−0.01	0.93
Serum HDL-C	0.05	0.26	−0.11	0.10	0.03	0.45	−0.12	0.05
Serum TAG	−0.10	0.05	−0.13	0.05	−0.07	0.10	−0.12	0.07
MetS[d]								
≥3 risk factors	−0.10	0.03	−0.08	0.22	−0.07	0.07	−0.05	0.35
MetS[d]	OR (CI)		OR (CI)		OR (CI)		OR (CI)	
(Reference = no MetS)	0.93 (0.80, 1.07)		0.93 (0.80, 1.07)		0.93 (0.80, 1.07)		0.93 (0.80, 1.07)	

MetS metabolic syndrome, *DP* dietary patterns, *PDP* prudent dietary pattern, *WDP* Western dietary pattern, *Std β* standardized β, *HOMA-IR* Homeostasis Model Assessment of insulin resistance, *QUICKI* Quantitative Insulin Sensitivity Check Index, *LDL-C* LDH cholesterol, *HDL-C* HDL cholesterol

HDL-C < 40 mg/dL in males and < 50 mg/dL in females; blood pressure ≥ 130 or ≥ 85 mm Hg or taking medications for hypertension; and fasting plasma Glucose ≥ 100 mg/dL or taking medications (oral hypoglycemic agents/insulin)

[a] Model 1 adjusted for age, energy intake, ethnicity, gender, ethnicity × gender, socio-economic status (SES), physical activity, alcohol intake and smoking status

[b] Model 2 adjusted for BMI in addition to age, energy intake, ethnicity, gender, ethnicity × gender, SES, physical activity, alcohol intake and smoking status

[c] BMI calculated as weight (kg)/height^2 (m^2) with normal weight defined as BMI ≥ 18.5 and ≤ 24.9 kg/m^2 and overweight/obese defined as BMI ≥ 25.0 g/m^2

[d] Diagnosis of MetS based on ≥ 3 of the following risk factors: waist circumference ≥ 102 cm in males and ≥ 88 cm in females; serum triglycerides ≥ 150 mg/dL

12.9 Gimme 5-A Fresh Nutrition Concept for High School Students

Data from the BHS indicated that improvements were needed in the diets of children. An effort was made to improve dietary habits of children. Nutrition is known to have a major impact on the development of certain cancers. [97–99]. According to Willet, "the inverse relationship between the intake of vegetables and fruit and the risk of cancer represents one of the best established associations in the field of nutritional epidemiology" [100]. Studies suggest that people who consume one or fewer daily servings of fruit and vegetables experience about twice the risk of developing cancer compared with those who consume four or more servings [101–103]. Based on the significant inverse association between the intake of vegetables and fruit and cancer risk, the National Cancer Institute (NCI) initiated the national 5-A-Day for Better Health Program in 1991 [104]. The goal of the campaign was to achieve one of the nation's health promotion and disease prevention objectives [105], a per capita intake of five servings of fruit and vegetables a day. Gimme 5: A Fresh Nutrition Concept for Students was one of nine NCI-funded research studies [106, 107] to evaluate population-based strategies to achieve the 5-A-Day goal. Gimme 5 focused on increasing fruit and vegetable consumption of high school students in the largest metropolitan area in Louisiana [104, 106–110].

12.9.1 Gimme 5 Intervention Program

The Gimme 5 program was designed to create an environment in which predisposing, enabling, and reinforcing factors described in the PRECEDE Model [110] positively affected daily consumption of fruit and vegetables. Consistent with the PRECEDE model, specific program components addressed the following levels of behavior change: Awareness Development, Interest Stimulation, Skills Training, Reinforcement, Application, and Maintenance.

Prior to program development, researchers conducted focus groups with high school students [107]. Focus group participants identified three barriers to increased fruit and vegetable consumption: (1) lack of availability; (2) lack of variety; and (3) inconsistency in taste. The intervention focused on these barriers.

Interventions comprised a school wide, media-marketing campaign; classroom activities; school meal modification ("Fresh Choices"); and parental involvement ("Raisin Teens"). Classroom and parent activities were delivered only to the intervention cohort; however, the entire school benefited from the school meal modification and media-marketing campaign. Details of the intervention were reported elsewhere [106, 107, 109], but an overview of the components follows.

12.9.1.1 Media-Marketing Campaign

A media-marketing campaign was the major intervention strategy for delivering 5-A-Day messages to students [111]. The campaign goal was to provide appealing messages and activities relevant to teenagers that would increase awareness, reinforce concepts, and promote positive attitudes toward consumption of fruit and vegetables. During the 1994–1995 school year, this intervention focused on monthly promotions of single fruit and vegetables with a theme and nutrition message. During the second and third year of the intervention, monthly promotions featured an ethnic theme.

Media materials and activities used in the monthly promotions included: (a) Marketing Stations: colorful three-paneled, six-foot vertical stations placed in the cafeterias to display information about Gimme 5 events, promotional materials, and concepts consistent with the 5-A-Day message; (b) Taste-Testings: monthly produce giveaways of fruit and vegetables and Gimme 5 recipes; (c) Point-of-Service Signs: colorful placards of monthly taste-testings of fruit and vegetables provided nutrient information per 5-A-Day serving size; (d) Posters: eye-catching posters displayed in high-visibility places, such as restrooms or locker rooms, to promote fruit and vegetable messages; (e) Table Tents: monthly table tents in the cafeteria featured 5-A-Day messages and activities about fruit and vegetables, encouraging student interaction and participation; (f) Public Service Announcements: schoolwide announcements were made informing students of Gimme 5 events, contest deadlines, and winnings; (g) Faculty Fruit and Vegetable Baskets: faculty fruit and vegetable baskets each semester encouraged consumption of a variety of fruit and vegetables; (h) Faculty Tip Sheets: monthly tip sheets distributed to faculty mailboxes provided nutritional information related to monthly promotions encouraging support and leadership for the Gimme 5 message; and, (i) Student Contests: activities conducted in the cafeteria to promote peer leadership and to stimulate student interaction.

12.9.1.2 Workshops

Each of the five, 55 min workshops contained a unique theme, a variety of learning strategies, and a student focus. Workshops provided students with learning opportunities to develop additional knowledge, positive attitudes, and skills necessary to increase fruit and vegetable consumption. Each workshop was designed to meet specific learning objectives related to the theme. Workshops were: (1) Fresh Start; (2) Body Works: Eating for Athletic Performance and Appearance; (3) Fresh Snax to the Max; (4) Fast Food-Go for the Green; and (5) Microwave Magic and Other Quick Fixes. Workshops were implemented by a Gimme 5 health educator or by school personnel selected by school principals and trained by Gimme 5 staff. The average length of training sessions for each workshop was about 50 min.

12.9.1.3 Supplementary Subject Activities

These activities were lessons included in required academic courses using fruit and vegetables in the lesson design. The purpose was to increase and maintain awareness by teachers and students of the 5-A-Day message during the first year of the intervention. Each teacher of the ninth grade cohort was requested to present at least one supplementary subject activity in their subject area each semester during the first year of intervention.

12.9.1.4 School Meal Modification

The goal of the school meal modification component, "Fresh Choices," was to increase the availability, variety, and taste of fruit and vegetables meeting 5-A-Day serving size and nutrient criteria in the school cafeteria. The component included four objectives: (1) increase student consumption of fruit and vegetables by increasing availability and portion size; (2) increase the variety of fruit and vegetables served; (3) increase acceptability and appeal of fruit and vegetable recipes to students; and (4) conduct monthly marketing activities to promote increased fruit and vegetable consumption.

Food service staff attended an initial training and follow-up booster sessions. Cycle menus developed by the school system were modified during the first year incorporating the monthly fruit and vegetable being promoted. During years two and three, 30 "Fresh Choices" guidelines were developed to assist cafeteria staff in modifying menus and recipes. Planners developed three ethnic menus for each ethnic promotion, with each cafeteria preparing a minimum of two of the three menus. Each menu contained at least three servings of fruit or vegetables. Entrees containing vegetables and fruit desserts were included as menu items, and proportioned salads and some cooked vegetables were increased on all menu days. Twenty-one ethnic menus were developed, incorporating 27 recipes that were developed and tested for acceptability among the high school students prior to implementation. During the last year of the intervention, cafeteria staff were encouraged to continue implementation of the "Fresh Choices" guidelines.

12.9.1.5 Parent Component

The goal of the parent component, "Raisin Teens," was to provide education, stimulate awareness, and elicit parental support for the Gimme 5 program. The objective was to encourage increased availability and variety of fruit and vegetables in the home. Gimme 5 staff conducted taste-testings of Gimme 5 recipes, media displays, and activities at Parent Teacher Organization meetings and at family-related functions. Colorful brochures featuring pictures of individual fruit and vegetables with recipes were distributed to parents with school mailings at least once a semester.

Brochures included purchasing tips, recipes, and nutritional information on individual fruit and vegetables that corresponded with the monthly promotions occurring at the school. The Gimme 5 Alive newsletter was sent each semester to parents of the intervention cohort. The newsletter provided information on Gimme 5 activities, recipes, discount coupons for produce, and information about the benefits and uses of fruit and vegetables. School newspapers and newsletters provided additional program information through a Gimme 5 column.

During the last year of the intervention, Gimme 5 intervention cohort parents received a calendar. The ethnic foods calendar provided monthly eating tips, and eating out and dishing it out (for home cooking) tips. The fifth day of each month was designated as Gimme 5 day to encourage home preparation of ethnic fruit and vegetable dishes. The calendar included coupons and recipe pages.

Gimme 5 provided a first model to show that dietary habits of high school students can be influenced by positive media messages relative to that age group, increased exposure to a variety of tasty products, and minimal classroom activity.

12.10 Summary

Diet is a major behavioral aspect related to health in diseases. The availability of foods and trends that have occurred greatly influence dietary intake. It is possible by careful dietary studies to document much of the adverse changes that influence risk factors in young individuals. The epidemic of obesity reflects trends in diet that have contributed to the dramatic occurrence of overweight and obesity. Through intensive media exposure it is possible to impose dietary consumption of fruit and vegetables as seen in a study of High School students and their parents.

Acknowledgments A special thank you to all of the contributing authors for the research included in this chapter: Mariam Morales, Adrianna Linares, Tom Baranowski, Carl DeMoor, Debby DeMory-Luce, Issa Zakeri, Su-Jau Yang, Nisha Mohindra, Summi Yoo, Priya Deshmukh-Taskar, Yan Liu, and Jeanette Gustat.

References

1. Merrill R (2010) Chapter 2: Historic development in epidemiology. Intorduction to epidemiology, 5th edn. Brigham Young University, Salt Lake City
2. Berenson GS, McMahan CA, Voors AW et al (1980) Cardiovascular risk factors in children: the early natural history of atherosclerosis and essential hypertension. Oxford University Press, New York
3. Berenson GS (1986) Causation of cardiovascular risk factors in children: perspectives on cardiovascular risk in early life. Raven Press, New York
4. Nicklas TA, Hayes D (2008) Position of the American Dietetic Association: nutrition guidance for healthy children ages 2 to 11 years. J Am Diet Assoc 108:1038–1044, 1046–1047
5. Farris RP, Nicklas TA (1993) Characterizing children's eating behavior. In: Suskind RM, Suskind LL, (eds) Textbook of pediatric nutrition, 2nd edn. New York, Raven Press. 2005–2016
6. Frank GC, Berenson GS, Schilling PE, Moore MC (1977) Adapting the 24 hour Recall for epidemiologic studies of school children. J Am Diet Assoc 71:26–31
7. The National School Lunch Program. http://www.fns.usda.gov/cnd/lunch. Accessed April 14, 2011
8. Nicklas TA, Farris RP, Bao W, Webber LS, Berenson GS (1997) Differences in reported dietary intake of 10-year-old children on weekdays compared to Sunday: the Bogalusa Heart Study. Nutr Res 17:31–40
9. Frank GC, Nicklas TA, Webber LS, Major C, Miller JF, Berenson GS (1992) A food frequency questionnaire for adolescents: defining eating patterns. J Am Diet Assoc 92:313–318

10. Rockett H, Breitenbach M, Frazier A et al (1997) Validation of a youth/adolescent food frequency questionnaire. Prev Med 26:808–816
11. Nicklas TA, Elkasabany A, Srinivasan SR, Berenson G (2001) Trends in nutrient intake of 10-year-old children over two decades (1973–1994): the Bogalusa Heart Study. Am J Epidemiol 153:969–977
12. Heart Disease and Stroke Statistics (2007) http://www.americanheart.org/downloadable/heart/1166712318459HS_StatsInsideText.pdf. Accessed April 24, 2011
13. Healthy People (2010) With understanding and improving health and objectives for improving health. Conference edition in two volumes 2nd edn. Washington: US Department of Health and Human Services, 2000
14. U.S. Department of Health and Human Services (1988) The surgeon general's report on nutrition and health. Washington: U.S. Government Printing Office
15. Weidman WJ, Kwiterovich P, Jesse MJ, Nugent E (1983) AHA committee report. Diet in the healthy child: task force committee of the nutrition committee and the cardiovascular disease in the young council of the american association. Circulation 67:1412A–1414A
16. National Academy of Sciences, National Research Council, Food and Nutrition Board (1989) Diet and Health: implications for reducing chronic disease risk. National Academy Press, Washington
17. Ahrens EH, Jr (1979) Introduction to symposium. The evidence relating six dietary factors to the nation's health: consensus statement. Am J Clin Nutr 32:2627–2631
18. American Academy of Pediatrics (1992) Committee on nutrition: statement on cholesterol. Pediatrics 90:469–473
19. Albertson A, Tobelmann R, Engstrom A (1992) Nutrient intakes of 2- to 10-year-old American children: 10-year trends. J Am Diet Assoc 12:1492–1496
20. Byers T (1993) Dietary trends in the United States. Relevance to cancer prevention. Cancer 72:1015–1018
21. Carroll MD, Abraham S, Dresser CM (1983) Dietary intake source data: United States, 1976–80. Vital & Health Statistics-Series II: data from the National Health Survey 11:1–483
22. Wilson JAW, Enns CS, Goldman JD (1998) Data tables: combined results from USDAs 1994 and 1995 continuing survey of food intakes of individuals (serial online)
23. Nicklas TA, Webber LS, Srinivasan SR, Berenson GS (1993) Secular trends in dietary intakes and cardiovascular risk factors of 10-year-old children: the Bogalusa Heart Study (1973–1988). Am J Clin Nutr 57:930–937
24. Nationwide Food Consumption Survey (1987) Continuing survey of food intakes by individuals. Hyattsville: Human Nutrition Information Services, Nutrition Monitoring Division, United States Department of Agriculture
25. Nationwide Food Consumption Survey (1988) Continuing survey of food intakes by individuals. Hyattsville: Human Nutrition Information Services. Nutrition Monitoring Division, United States Department of Agriculture
26. Alaimo K, McDowell MA, Briefel RR et al (1994) Dietary intake of vitamins, minerals and fiber of persons ages 2 months and over in the United States: third National Health and Nutrition Examination Survey, Phase 1, 1988–91. National Center for Health Statistics, Hyattsville
27. Centers for Disease Control and Prevention NHANES I (2010) In: Statistics NCfH (ed) National Health and Nutrition Examination Survey. Atlanta
28. Nicklas TA, Morales M, Linares A et al (2004) Children's meal patterns have changed over a 21-year period: the Bogalusa Heart Study. J Am Diet Assoc 104:753–761
29. Ceru-Bjork C, Andersson I, Rossner S (2001) Night eating and nocturnal eating-two different or similar syndromes among obese patients? Int J Obes Relat Metab Disord 25:365–372
30. Speechly DP, Rogers GG, Buffenstein R (1999) Acute appetite reduction associated with an increased frequency of eating in obese males. Int J Obes Relat Metab Disord 23:1151–1159
31. Fabry P, Fodo J, Hejl Z, Braun T, Zvolankova K (1964) The frequency of meals: its relation to overweight, hypercholesterolaemia, and decreased glucose tolerance. Lancet 2:614–615

32. Kant AK, Schatzkin A, Graubard BI, Ballard-Barbash R (1995) Frequency of eating occasions and weight change in the NHANES I epidemiologic follow-up study. Int J Obes Relat Metab Disord 19:468–474
33. Metzner HL, Lamphiear DE, Wheeler NC, Larkin FA (1977) The relationship between frequency of eating and adiposity in adult men and women in the Tecumseh Community Health Study. Am J Clin Nutr 30:712–715
34. Drewnowski A, Cohen AE, Faust IM, Grinker JA (1984) Meal-taking behavior is related to predisposition to dietary obesity in the rat. Physiol Behav 32:61–67
35. Edelstein SL, Barrett-Connor EL, Wingard DL, Cohn BA (1992) Increased meal frequency associated with decreased cholesterol concentrations; Rancho Bernardo, CA, 1984–1987. Am J Clin Nutr 55:664–669
36. Gwinup G, Byron RC, Roush WH, Kruger FA, Hamwi GJ (1963) Effect of nibbling versus gorging on serum lipids in man. Am J Clin Nutr 13:209–213
37. Jenkins DJ, Wolever TM, Vuksan V et al (1989) Nibbling versus gorging: metabolic advantages of increased meal frequency. N Engl J Med 321:929–934
38. Nunes WT, Canham JE (1963) The effect of varied periodicity of eating on serum lipids and carbohydrate tolerance in man. Am J Clin Nutr 12:334
39. Cohn C (1961) Meal-eating, nibbling and body metabolism. J Am Diet Assoc 38:433
40. Arnold LM, Ball MJ, Duncan AW, Mann J (1993) Effect of isoenergetic intake of three or nine meals on plasma lipoproteins and glucose metabolism. Am J Clin Nutr 57:446–451
41. McGrath SA, Gibney MJ (1994) The effects of altered frequency of eating on plasma lipids in free- living healthy males on normal self-selected diets. Eur J Clin Nutr 48:402–407
42. Mann J (1997) Meal frequency and plasma lipids and lipoproteins. Br J Nutr 77:S83–S90
43. Bertelsen J, Christiansen C, Thomsen C et al (1993) Effect of meal frequency on blood glucose, insulin, and free fatty acids in NIDDM subjects. Diabetes Care 16:4–7
44. Jones PJ, Leitch CA, Pederson RA (1993) Meal-frequency effects on plasma hormone concentrations and cholesterol synthesis in humans. Am J Clin Nutr 57:868–874
45. Garcia SE, Kaiser LL, Dewey KG (1990) The relationship of eating frequency and caloric censity to energy intake among rural Mexican preschool children. Eur J Clin Nutr 44:381–387
46. Verboeket-van de Venne WP, Westerterp KR (1991) Influence of the feeding frequency on nutrient utilization in man: consequences for energy metabolism. Eur J Clin Nutr 45:161–169
47. Nicklas TA, Demory-Luce D, Yang S-J, Baranowski T, Zakeri I, Berenson G (2004) Childrens food consumption patterns have changed over two decades (1973–1994): the Bogalusa Heart Study. J Am Diet Assoc 104:1127–1140
48. Kennedy E, Powell R (1997) Changing eating patterns of American children: a view from 1996. J Am Diet Assoc 16:524–529
49. Johnson RK (2000) Changing eating and physical activity patterns of US children. Proc Nutr Soc. 59:295–301
50. Demory-Luce D, Morales M, Nicklas T, Baranowski T, Zakeri I, Berenson G (2004) Changes in food group consumption patterns from childhood to young adulthood: the Bogalusa Heart Study. J Am Diet Assoc 104:1684–1691
51. Cavadini C, Siega-Riz AM, Popkin BM (2000) US adolescent food intake trends from 1965 to 1996. Arch Dis Child 83:18–24
52. Nicklas TA, Yang SJ, Baranowski T, Zakeri I, Berenson G (2003) Eating patterns and obesity in children: the Bogalusa Heart Study. Am J Prev Med 25:9–16
53. Mohindra NA, Nicklas TA, O'Neil CE, Yang SJ, Berenson GS (2009) Eating patterns and overweight status in young adults: the Bogalusa Heart Study. Int J Food Sci Nutr 60 (Suppl 3):14–25
54. Flegal KM (1999) The obesity epidemic in children and adults: current evidence and research issues. Med Sci Sports Exerc 31:S509–S514
55. Rippe JM (1998) The obesity epidemic: challenges and opportunities. J Am Diet Assoc 98:S5
56. Ogden CL, Carroll MD, Curtin LR, Lamb MM, Flegal KM (2007–2008) Prevalence of high body mass index in US children and adolescents, 2007–2008. J Am Med Assoc 303:242–249

57. U.S. Department of Agriculture (1998) Childhood obesity: causes and prevention. Proceedings of a symposium sponsored by the Center for Nutrition Policy and Promotion, USDA. October 27, 1998. Available at http://www.nal.usda.gov:8001/Training/obesity.pdf. Accessed July 2002. Washington
58. Zemel MB, Shi H, Greer B, Dirienzo D, Zemel PC (2000) Regulation of adiposity by dietary calcium. Fed Am Soc Exp Biol J 14:1132–1138
59. Roberts SB, McCrory MA, Saltzman E (2002) The influence of dietary composition on energy intake and body weight. J Am Coll Nutr 21:140S–145S
60. Gibson SA (1996) Are high-fat, high-sugar foods and diets conducive to obesity? Int J Food Sci Nutr 47:405–415
61. Macdiarmid JI, Vail A, Cade JE, Blundell JE (1998) The sugar-fat relationship revisited: differences in consumption between men and women of varying BMI. Int J Obes Relat Metab Disord 22:1053–1061
62. Bray GA, Popkin BM (1998) Dietary fat intake does affect obesity! Am J Clin Nutr 68:1157–1173
63. Mancino M, Todd J, Guthrie J, Lin B-H (2010) How food away from home affects children's diet quality. Economic Research Report No. (ERR-104)
64. Clemens LH, Slawson DL, Klesges RC (1999) The effect of eating out on quality of diet in premenopausal women. J Am Diet Assoc 99:442–444
65. McCrory MA, Fuss PJ, McCallum JE et al (1999) Dietary variety within food groups: association with energy intake and body fatness in men and women. Am J Clin Nutr 69:440–447
66. Coll M, Meyer A, Stunkard AJ (1979) Obesity and food choices in public places. Arch Gen Psychiatry 36:795–797
67. Ludwig DS, Peterson KE, Gortmaker SL (2001) Relation between consumption of sugar-sweetened drinks and childhood obesity: a prospective, observational analysis. Lancet 57:505–508
68. Dennison BA, Rockwell HL, Baker SL (1997) Excess fruit juice consumption by preschool-aged children is associated with short stature and obesity. Pediatrics 99:15–22
69. Hill JO, Peters JC (1998) Environmental contributions to the obesity epidemic. Science 280:1371–1374
70. Basiotis PP, Lino M, Anand RS (1998) Report card on the diet quality of african Americans. Family Econ Nutr Rev 11:61–63
71. Forshee RA, Storey ML (2004) Controversy and statistical issues in the use of nutrient densities in assessing diet quality. J Nutr 134:2733–2737
72. Bachman CM, Baranowski T, Nicklas TA (2006) Is there an association between sweetened beverages and adiposity? Nutr Rev 64:153–174
73. Harnack L, Stang J, Story M (1999) Soft drink consumption among US children and adolescents: nutritional consequences. J Am Diet Assoc 99:436–441
74. Yoo S, Nicklas T, Baranowski T et al (2004) Comparison of dietary intakes associated with metabolic syndrome risk factors in young adults: the Bogalusa Heart Study. Am J Clin Nutr 80:841–848
75. Reaven GM (1998) Role of insulin resistance in human disease. Diabetes 37:1595–1607
76. Ford ES, Li C, Zhao G (2010) Prevalence and correlates of metabolic syndrome based on a harmonious definition among adults in the US. J Diabetes 2:180–193
77. Singh AS, Mulder C, Twisk JW, van Mechelen W, Chinapaw MJ (2008) Tracking of childhood overweight into adulthood: a systematic review of the literature. Obes Rev 9:474–488
78. Chen X, Wang Y (2008) Tracking of blood pressure from childhood to adulthood: a systematic review and meta-regression analysis. Circulation 117:3171–3180
79. Webber LS, Srinivasan SR, Wattigney WA, Berenson GS (1991) Tracking of serum lipids and lipoproteins from childhood to adulthood: the Bogalusa Heart Study. Am J Epidemiol 133:884–899
80. Pagnan A, Ambrosio GB, Vincenzi M et al (1982) Precursors of atherosclerosis in children: the cittadella study. Follow-up and tracking of total serum cholesterol, triglycerides, and blood glucose. Prev Med 11:381–390

81. Zimmet P, Alberti G, Kaufman F et al (2007) The metabolic syndrome in children and adolescents. Lancet 369:2059–2061
82. Reaven GM (2000) Diet and syndrome X. Curr Atheroscler Rep 2:503–507
83. Abbasi F, McLaughlin T, Lamendola C et al (2000) High carbohydrate diets, triglyceride-rich lipoproteins, and coronary heart disease risk. Am J Cardiol 85:45–48
84. Wirfalt E, Hedblad B, Gullberg B et al (2001) Food patterns and components of the metabolic syndrome in men and women: a cross-sectional study within the malmo diet and cancer cohort. Am J Epidemiol 154:1150–1159
85. Ludwig DS, Pereira MA, Kroenke CH et al (1999) Dietary fiber, weight gain, and cardiovascular disease risk factors in young adults. J Am Med Assoc 282:1539–1546
86. Pereira MA, Jacobs DR Jr, Van Horn L, Slattery ML, Kartashov AI, Ludwig DS (2002) Dairy consumption, obesity, and the insulin resistance syndrome in young adults: the CARDIA study. J Am Med Assoc 287:2081–2089
87. Fung TT, Rimm EB, Spiegelman D et al (2001) Association between dietary patterns and plasma biomarkers of obesity and cardiovascular disease risk. Am J Clin Nutr 73:61–67
88. Miyatake N, Nishikawa H, Morishita A et al (2002) Daily walking reduces visceral adipose tissue areas and improves insulin resistance in Japanese obese subjects. Diabetes Res Clin Pract 58:101–107
89. Meigs JB, D'Agostino RB Sr, Wilson PW, Cupples LA, Nathan DM, Singer DE (1997) Risk variable clustering in the insulin resistance syndrome. The framingham offspring study. Diabetes 46:1594–1600
90. McAuley KA, Williams SM, Mann JI et al (2002) Intensive lifestyle changes are necessary to improve insulin sensitivity: a randomized controlled trial. Diabetes Care 25:445–452
91. Jacques PF, Tucker KL (2001) Are dietary patterns useful for understanding the role of diet in chronic disease? Am J Clin Nutr 73:1–2
92. Deshmukh-Taskar PR, O'Neil CE, Nicklas TA et al (2009) Dietary patterns associated with metabolic syndrome, sociodemographic and lifestyle factors in young adults: the Bogalusa Heart Study. Public Health Nutr 12:2493–2503
93. Hu FB (2002) Dietary pattern analysis: a new direction in nutritional epidemiology. Curr Opin Lipidol 13:3–9
94. Willett W (1990) Nutritional epidemiology. Oxford University Press, New York
95. Freeland-Graves J, Nitzke S (2002) Position of the American Dietetic Association: total diet approach to communicating food and nutrition information. J Am Diet Assoc 102:100–108
96. Brunner EJ, Mosdol A, Witte DR et al (2008) Dietary patterns and 15-y risks of major coronary events, diabetes, and mortality. Am J Clin Nutr 87:1414–1421
97. Key TJ (2011) Fruit and vegetables and cancer risk. Br J Cancer 104:6–11
98. Wallström P, Wirfält E, Janzon L, Mattisson I, Elmståhl S, Johansson U, Berglund G (2000) Fruit and vegetable consumption in relation to risk factors for cancer: a report from the Malmö Diet and Cancer Study. Public Health Nutr 3:263–271
99. Chen VW, Fontham E, Groves FD, Craig JF, Correa P (1991) Cancer incidence in south Louisiana: 1983–1986. Cancer In Louisiana VII
100. Willett WC (1990) Vitamin A and lung cancer. Nutr Rev 48:201–211
101. Block G, Patterson B, Subar A (1992) Fruit, vegetables, and cancer prevention: a review of the epidemiological evidence. Nutr Cancer 18:1–29
102. Nationwide Food Consumption Survey (1986) Continuing survey of food intakes by individuals: women 19–50 years and their children 105 years, 1 day. In: Service HNI, (ed) Report No. 86–1. Washington: US Department of Agriculture
103. Patterson B, Block G (1991) Fruit and vegetable consumption: national survey data. In: Bendich E, Butterworth C (eds) Micronutrients in health and the prevention of disease. Dekker, New York 409–436
104. Produce for better health foundation and National Cancer Institute (1991) 5-a-day for better health guidebook

105. Healthy People (1991) 2000: National health promotion and disease prevention objectives. Washington: United States Department of Health and Human Services. Public Health Serv, pp.101–103
106. Havas S, Heimendinger J, Reynolds K et al (1994) 5 a day for better health: a new research initiative. J Am Diet Assoc 94:32–36
107. Havas S, Heimindinger J, Damron D et al (1995) 5-a-day for better health—nine community research projects to increase fruit and vegetable consumption. Public Health Rep 110:68–79
108. Beech BM, Rice R, Myers L, Johnson C, Nicklas TA (1999) Knowledge, attitudes, and practices related to fruit and vegetable consumption of high school students. J Adolesc Health 24:244–250
109. Nicklas TA, Johnson CC, Farris RP, Rice RR, Lyon L, Shi R (1997) Development of a school-based nutrition intervention for high school students: gimme 5. Am J Health Promot 11:315–322
110. Green LW, Kreuter M, Deeds SG, Partridge KB (1980) Health education planning: a diagnostic approach. Mayfield, Palo Alto
111. Nicklas TA, Johnson CC, Myers L, Webber LS, O'Neil C (2000) Using a media-marketing campaign for promoting increased awareness about fruit and vegetable consumption among high school students: gimme 5 program. J Child Nutr Manag 24:27–34

Chapter 13
Cardiovascular Health Promotion—Physical Fitness in the School Setting

Marietta Orlowski, James Ebert and Arthur Pickoff

Abstract Physical activity is a major lifestyle behavior related to cardiovascular disease. Lack of energy expenditure is considered an important contributor to the epidemic of obesity. Schools are an obvious area of interest in health promotion due to their centrality in the life of children. Over the course of the school day, physical education, recess, extracurricular activities, and classroom activities provide multiple opportunities to integrate and reinforce physical activity. School policies exert a powerful influence in limiting and promoting these health and fitness opportunities. A high level of interest in youth fitness began in the 1950s and continues today. Yet these school day activity opportunities are rife with policy contractions. This chapter reviews the role of schools in the promotion of physical activity and fitness.

Keywords Physical activity • Early childhood • Health promotion • Fitness • Physical education

13.1 Introduction

The role of physical activity and nutrition in the promotion of child and adolescent cardiovascular health is well established [37]. In early childhood, nearly all instruction related to exercise and nutrition takes place in the home and is provided by parents or guardians. During pediatric health supervision visits, anticipatory guidance emphasizes providing parents with knowledge and skills for health promotion [13]. As children enter the education system and spend less time at home, parents begin sharing the role of teaching and mentoring their children with schools. The amount of time spent in school settings increases as they progress though childhood and adolescence.

M. Orlowski (✉)
Department of Community Health, Boonshoft School of Medicine,
Wright State University, Dayton, OH, USA
e-mail: marietta.orlowski@wright.edu

G. S. Berenson (ed.), *Evolution of Cardio-Metabolic Risk from Birth to Middle Age*,
DOI 10.1007/978-94-007-1451-9_13, © Springer Science+Business Media B.V. 2011

The relationship of one's environment with physical activity and nutrition behaviors is also well established. Individuals exist in physical and social environments, and these environments influence the foods one chooses to consume, and the sedentary and active behaviors one may engage in. In this ecological framework, environments exert influence on a multitude of levels [34]. The individual (intrapersonal) is influenced by the social and cultural practice of family and friends (interpersonal) as well as the environments of the organizations and neighborhoods of daily living. Neighborhood factors include recreational facilities, grocery stores, transportation options, and neighborhood design. Individuals and communities also exist in larger eco-systems, where decisions are far removed from the individual, and these systems can have tremendous influence on daily behaviors [30]. The price and accessibility of products, such as tobacco or fast food, are two such examples.

Schools are an obvious area of interest in health promotion due to their centrality in the life of children and the potential availability of equipment, facilities, and facilitators. School attendance is mandatory until at least age 16 in all states; and the length of the school year varies by state, but ranges from 175 to 185 days [11]. Schools also have the effect of creating *de facto* cohorted community subgroups consisting of the parents of enrolled children and education professionals. This occurs in all traditional school settings, whether the community of families is defined by a public school district's geographic boundaries or by the parental choice of private or faith-based education. This cohorting of parents through their children's school provides an opportunity for schools to influence parenting in the home, thus indirectly benefiting the students, to complement the education that students receive directly on campus. Since nearly all schools are either intrinsically the instruments of government (public schools) or subject to regulation and accreditation standards (private, including faith-based schools), government agencies can readily target them for the purpose of providing resources, or imposing standards in the interest of promoting health. The institutions of education thus provide the best means for reaching the majority of school-age children with population-based health promotion and fitness interventions.

Policies are powerful tools in promoting student health and fitness. The power of policies comes from creating environments that support individual behaviors. Without supportive physical and social environments, individual behavior changes are difficult to sustain. The school environment represents numerous built and social factors external to the student and staff: walkways to and from school, playgrounds, classroom space, equipment, after school programs, vending machines, and à la carte foods. Accordingly, this chapter will review the role of schools in the promotion of physical activity and fitness, with an emphasis on population-level environmental and procedural practices. A brief historical review will provide context for school-based changes in policy and practice needed to promote better fitness.

13.2 School Day Physical Activity Opportunities

Over the course of a school day, there are multiple opportunities to integrate and reinforce physical activity. Ideally, the combination of physical education, recess, extracurricular activities, and classroom activities would ensure that children meet the minimum recommendation of 60 min a day of physical activity per day [8, 15, 22].

13.2.1 Walking to School

A school day of fitness opportunities begins with the way a child travels to and from school. Approximately 15% of students walk to and from school [47]. Walking rates vary by neighborhood and walker characteristics. Distance from school is consistently the most influential predictor of mode of transportation. Students who reside within one mile of the school are three times more likely to walk or bicycle to school [1, 7]. The condition of the route to school, and one's perception of route safety, also influence the rates of active transportation. Routes with functional sidewalks, crosswalks, trees and other ascetics, fewer vacant lots, and calm traffic are considered more "walkable" [1, 12]. Individually, students under age 14 are more likely to walk or bicycle [16], and when gender differences exist, males are more likely to walk or bicycle to school [5, 32].

Arriving at school via active means was more common in the previous decades. In 1969, approximately half of all students walked or bicycled to or from school [46]. A notable portion of the decrease in active commuters is attributed to the changes in home and school proximity. The suburban sprawl of the 1950s and 1960s was accompanied by the strengthening of an automobile-dependant life style [47]. The design of neighborhoods changed accordingly. Rural areas near cities became transformed to new suburbs which often lacked sidewalks, and children gathered at bus pick-up points for what had become a longer ride to school. In urban communities, federal mandates for racial integration of public schools had similar effects on home and school proximity. Brown vs. the Board of Education (1954) ushered in an era of policies to address racial integration of public schools. "Forced busing" of the 1970s was implemented in numerous districts in which students could be transported to a school miles outside their neighborhood [33]. As a result of sprawl, school construction, integration, and school innovations such as charter or magnet schools, half of all students now live three or more miles from their school [47].

Distance is not the sole moderator of active commuting to and from school. Current local programs to promote walking and biking to school are encouraged, and possibly funded, by federal programs like the Safe Routes to School, administered by the US Department of Transportation's Federal Highway Administration. A 5 E's model of Engineering, Enforcement, Education, Encouragement and Evaluation funds policy development and enforcement, and infrastructure changes, along with parent and child educational programs to alter beliefs about safety, and reinforcement of the active behavior. To date 5,464 programs, across all 50 states and the District of Columbia, have been funded [24].

13.2.2 Physical Education

Structured physical activity at school is most visibly associated with physical education. Physical education is defined as the curricular area, taught by professionals, which develops the skills and knowledge to establish and maintain an active lifestyle [23]. Currently, 90% of school districts require physical education; however, the time and content requirements vary greatly from state to state, and by grade level [23]. Physical education (PE) is most often mandated for high school students (90%) as a requirement for graduation and mandates typically range from 0.5 credits to 2 credits. New Jersey, a true outlier, requires the equivalent of 15 credits [23]. In the lower grades, 84% of schools require PE in elementary grades, while 76% of middle school students are required to participate in PE. About one-third of these districts specify minutes per week of physical education: three states require the national recommended amount: 150 min per week in elementary and 225 min per week in middle school [23]. Programs of academic instruction, or scope and sequence, are guided by academic content standards. Most states (92%) have state physical education standards, however, only 67% of states require local districts to comply with these standards. Thus, instruction quality and minutes of activity varies by state, school, and possibly building. Furthermore, assessment, a mechanism to aggregate and report student performance, is not as standardized as other content areas like science, mathematics and reading. One-third of states require assessment of student PE performance or fitness, and only 10% report his information to a state department of education [23].

High level interest in youth fitness became evident as early as the 1950s when President Eisenhower created the President's Council on Youth Fitness, a cabinet-level office responsible for implementing a nationwide pilot study of the fitness level of 8,500 boys and girls ages 5 through 12. This study resulted in a national testing program, known today as the President's Challenge. President Kennedy called for the involvement of all citizens and community groups to promote fitness not just of youth, but of all Americans, and began personally participating in 50 mile hikes to demonstrate his commitment to the program [38, 48]. Under Kennedy's administration, pilot fitness programs in schools across the nation were launched. President Johnson later administratively placed the Council on Youth and Fitness under the Department of Health Education and Welfare (now the Department of Health and Human Services), and every president and every Congress since that time have issued proclamations and or initiatives in support of youth fitness [38, 48]. Two contemporary campaigns, "Let's Move", a program initiated by First Lady Michele Obama, and "Alliance for a Healthier Generation" co-created by former President William J. Clinton have missions of raising a healthier generation through physical activity and dietary changes.

Physical education in schools is rife with policy contradictions. The attention and discussion on youth physical activity is intense, due in large part to the growing problem of obesity. Despite that national concern, physical education is not considered an academic subject, and thus, resources are compromised to accommodate the academic content pressures and mandates. For example, 63% of school districts allow students to substitute, or be exempted from physical education. These percentages are up significantly from 2006 in which 53% of districts allowed substitutions, and 35% allowed waivers [23]. Similarly, in the 2001 landmark legislation No Child Left Behind (NCLB), physical education was ignored as an academic content area [23]. Also in 2001, the Carol M. White Physical Education Program (PEP) was established by the US Department of Education to fund grant applications for schools and local education agencies to improve physical education and activity programs. This one-time award money provides updated equipment and some professional development, but PEP does not address the systemic constraints holding PE back—qualified teachers, quality of instruction, quantity of time, and number of attending students [35]. Lastly, the US Department of Transportation spent $612 million between 2005 and 2010 promoting walking and bicycling to school, yet one-third of elementary students go to school in a district which prohibits walking to school [43].

13.2.3 Organized Extracurricular Activities and Sport

After-school sports and extracurricular activities provide additional opportunities for physical activity. At least half of all high school students participate in an organized sport [6, 16, 26]. High school students are more likely to participate in an interscholastic or varsity sports (33–37%), compared with 16–20% of high school students participating in intramural or a physical activity club [16, 20]. Current participation rates do not vary by gender, but sport preferences do vary. Football, track and field, basketball, baseball, soccer, wrestling, and cross-country are the most popular programs for males. Track and field, basketball, volleyball, softball, soccer, cross country, and tennis are the most popular programs for females [26].

Participation trends vary by competitive level, which can be dichotomized into varsity sports and intramural sport. The number of students participating in competitive varsity sports is either stable [16] or increasing [26]. In contrast, involvement in intramurals and physical activity clubs declines as students progress through school. About one in five boys and girls, 24.0% and 20.5%, respectively, participates in an intramural sport or club; by graduation, participation drops to one in seven students (15.8% and 13.3%).

Title IX of the Education Amendments Act of 1972 has had a significant impact on the participation by girls in both community and school-based sports. Title IX prohibits sex discrimination in any educational program or activity by schools receiving federal financial assistance [44]. In athletics, the federal law requires equitable opportunity to participate in sports; equal opportunity for financial scholarships; and similar provisions such as equipment, facilities, and coaching [50]. As a result of Title IX, high school participation has increased by 904% [50]. Prior to Title IX, the primary physical activities for girls were cheerleading and square-dancing. There were virtually no college scholarships for female athletes. Only 2% of overall athletic budgets went to female college athletes [39].

Organized sports and recreation can provide moderate to vigorous physical activity. Interscholastic varsity sports appear to be resistant to funding shortfalls in many districts as evidenced by stable to increasing numbers of student athletes. A reemphasis on intramurals and club sports might engage students who are less athletically inclined. Intramural environments can be less competitive, more enjoyable, and inclusion of varying skill levels [19].

13.2.4 Recess and Play

Playground areas have been commonplace on elementary school campuses throughout the past six decades. In the grades before middle school, children are provided varying amounts of activity time known as recess. Daily recess provides the largest opportunity, followed by physical education, for all students to be physically active [4, 29, 42]. Recess policies vary by school and student characteristics; one national report estimated average recess ranged from 27.8 min for first grade students to 23.8 min for sixth graders [45]. Similarly, in a Gallup Poll of elementary school principals, half of surveyed principals reported a daily recess of 16 to 30 min per day [31]. Urban schools, schools with a majority of ethnically minority students, and large schools tend to report lower minutes and offerings of recess [3, 27]. Recess is supervised, but generally unstructured in play [29]. Thus, the extent to which children are active during recess varies. Gender activity level differences have been reported, but findings are inconsistent. When differences do exist, boys tend to engage in more moderate to vigorous activities than girls during the allotted recess time [21, 28]. Recess activity levels also vary by school and playground characteristics. Active supervision, playground facilities, and portable play equipment availability can all influence student population activity levels [41, 49].

The future of recess is teetering. On a local level, elementary school principals and parents overwhelming support daily recess [31]. Recess has been associated with better classroom behavior and social development [3, 8]. Furthermore, eight out of ten principals report perceived academic benefits, yet one of five reported decreases in allotted recess time due to annual yearly progress testing [31]. Standardized academic testing and other time-intensive mandates leave recess vulnerable. Recess champions have defended recess as an underutilized opportunity for activity [3, 29, 31]. Advocacy groups have encouraged districts, schools and parents to stand up for recess. They encourage fun activities, making age-appropriate equipment available and recruiting parents and volunteers as facilitators [2, 29].

13.3 Classroom Integration

Students spend the greatest portion of their school day in the academic classroom with numerous opportunities for physical activity. Up to 25 h a week can be spent in a classroom, compared with an average of one hour a week in physical education [14, 25]. Teachers can incorporate physical activity via games, integrated lessons, and into routines and transitions. Games—the traditional way to offer physical activity in a classroom—are separate events provided as a break from learning content and completing work. Physical activity can also be integrated into content lessons. For example, in a mathematics lesson, children can wear pedometers, circulate through calisthenic stations, then calculate ranges and averages from the number of recorded steps or chart the total [36]. Finally, physical activity can be incorporated into daily routines and transitions. Typically, a class follows routines around the daily schedule, such as a consistent set of activities each morning. Transitions are times in the day when children change from one space or subject to a new one. Students transition to move outdoors, eat, wash hands, or go from reading circle to team tables or from art to reading. Movement during routines and transitions requires only personal space and no additional resources.

Preliminary studies document various positive physiological, cognitive, and emotional benefits even with moderate activity [8]. Physiologically, movement expends energy and increases blood flow; cognitively, brief activities improve attention, memory, and learning; and emotionally, brief activities improve mood [8, 40]. Movement throughout of the day also helps students and teachers establish activity routines that can be transferred to the more challenging contexts, such as indoor recess. Lastly, taking part in simple movement activities throughout the day reinforces the beliefs, skills, and behaviors of a physically active lifestyle.

Incorporating physical activity into the routines and learning activities of the classroom is in the early stages of system integration and evaluation. No federal mandates specifically address, nor fund, this type of school-based physical activity. A growing but still limited number of states have mandated daily activity minimums. Classroom integration could address such mandates, and benefit students with no additional requirements for resources. In the School Health Policies and Programs Study of the CDC, 15.5% of surveyed districts required elementary schools to provide regular physical activity breaks in addition to recess and physical education [18]. Mandates decrease as student grade levels increase: 10.0% of surveyed middle schools districts and 3.8% of surveyed high schools required physical activity breaks.

13.4 Local School Wellness Policies

Future school wellness actions will certainly be shaped by the federal Child Nutrition and WIC Reauthorization Act of 2004. As of 2006, all public and private schools receiving federal school meal funding are required to have school wellness policies that aim to "promote student wellness" and "reduce childhood obesity". The federal policy is not content specific, serving to encourage local agencies to tailor activities to school and community need. Minimally, school wellness policies should address the following: (1) goals for nutrition education, physical activity and other school-based activities designed to promote student wellness; (2) nutrition guidelines for all foods available on each school campus with the objectives of promoting student health and reducing childhood obesity; (3) guidelines for reimbursable school meals, which are no less restrictive than existing federal regulations; (4) a plan for measuring implementation of the local wellness policy, including designation of one or more persons charged with operational responsibility for ensuring that each school fulfills the district's local wellness policy and (5) community involvement in the development of the school wellness policy. Sample physical activity school policies include daily recess requirements, physical education teacher qualifications, busing of students, and the use of physical activity as a punishment.

Energy expenditure during various activities is balanced by energy and nutritional intake. Ideally, school breakfast, lunch, and à la carte food offerings also provide opportunities for health enhancing behaviors. Sample school nutrition wellness policies include restrictions on soda and sugar sweetened beverages, restrictions on food served at classroom parties, and limits on fat and calorie content of à la carte foods [2, 10].

The potential for wellness policies to create health-enhancing environments is tremendous; however, the effective use of these policies is in the early stages. Two recent nationwide studies concluded that a high percentage of schools have written wellness policies, but that the strength of the policies is low and the effectiveness of the policies is not measured [9, 10]. For example, in a review of the strength of written school wellness policy, Chriqui et al. [9] found the average scores of strength ranged from 29 to 35 points on a 100-point scale. Researchers found underdeveloped and fragmented policies, and insufficient plans for monitoring and evaluating policy [9, 10]. The strongest policies, and the ones most likely to be implemented, were policies that restricted school foods and beverages [9].

13.5 For the Future—A Coordinated Approach

Future solutions for increasing physical activity levels of children will be based in a coordinated school approach. A coordinated approach identifies multiple avenues to integrate the target behavior and design strategies that can work synergistically. Over the course of a school day, there are multiple opportunities to integrate, and reinforce, physical activity. No single strategy, such as physical education or recess, can provide the skill development, reinforcement, and time needed to shape and sustain the necessary health-related behaviors.

School-based strategies should also be coordinated and evaluated around specific defined health behaviors. Physical activity goals may be as simple as minutes of physical activity in a school day or as specific as bouts of moderate to vigorous physical activities. Fragmented programs and evaluations tend to be designed and evaluated based on activities or anthropometrics, instead of behaviors. Body Mass Index (BMI) has become an internationally accepted metric for obesity screening [13], and has also become the most common evaluation measure reported for school wellness policies [9]. BMI is an appropriate screening measure of childhood weight and obesity, and is easy to measure and monitor. However, BMI is not the outcome of a single intervention, and is an unlikely annual measure of program or policy effectiveness [9].

One practical model of coordination and behavioral focus is the Coordinated School Health Program (CSHP), developed by the Centers for Disease Control and Prevention. The CSHP identifies specific disciplines and community members that can work collaboratively to promote all student and staff wellness objectives. The eight-component model includes physical education, health education, school health services nutrition services, counseling services, school policies and environment, health promotion for staff and community and family involvement [17]. Table 13.1 demonstrates how each of these components serves a role in promoting the activity and fitness level of school children.

Table 13.1 Physical activity in a coordinated school health program. [17]

School component	Description	Sample physical activity policies
Health education	A planned, sequential, K-12 curriculum that addresses health-related knowledge, attitudes, skills, and practices related to the physical, mental, emotional and social dimensions of health	Licensed health education teachers Standards-based curriculum with behavioral nutrition and physical activity goals
Physical education	A planned, sequential K-12 curriculum that provides cognitive content and learning experiences in a variety of activity areas such as basic movement skills; physical fitness; rhythms and dance; games; team, and individual sports; tumbling and gymnastics; and aquatics	Licensed PE Teachers Pupil to teacher ratios Assessment of student fitness 150 min weekly PE elementary grades 225 min PE middle and secondary grades
School health services	Services designed to ensure access or referral to primary health care services, foster appropriate use of primary health care services, prevent and control communicable disease and other health problems, and provide emergency care for illness or injury	School health advisory committee Body mass index (BMI) screenings
Nutrition services	Access to a variety of nutritious and appealing meals and food choices that accommodate the health and nutrition needs of all students	Nutrition standards for à la carte foods and vending machines Guidelines for food at extracurricular events and classroom parties

Table 13.1 (continued)

School component	Description	Sample physical activity policies
Counseling services	Individual and group assessments, interventions, and referrals provided to improve students' mental, emotional, and social health. Professionals such as certified school counselors, psychologists, and social workers provide these services	Monitoring of student PE attendance Body mass index (BMI) screenings counseling
School policy and environment	The physical and aesthetic surroundings and the psychosocial climate and culture of the school. The psychological environment includes the physical, emotional, and social conditions that affect the wellbeing of students and staff	Provision of equipment for indoor and outdoor recess Safe walking/bicycling routes to school
Health promotion for staff	Opportunities for school staff to improve their health status and health-related behaviors. A personal commitment often transfers into greater commitment to the health of students and creates positive role modeling	Incentives for staff health promoting behavior
Community and family involvement	Connecting school activities to outside resources and opportunities as well as connecting home-schooled students to community-based physical activities	Open facilities after school hours for use by students and family Recreation leagues for students of all abilities Monitoring of walking routes to and from school

13.6 Conclusions

Schools present a logical opportunity for health promotion, due to their centrality in the life of children and adolescents. Just as school components must work in a coordinated fashion, so must the interventions of federal, state, and local agencies. Past and current federal government initiatives and mandates have captured the spirit of health and fitness, but implementation and evaluation will remain the responsibility of local and regional school authorities.

References

1. Active Living Research (ALR) (2009, May) Walking and biking to school, physical activity and health outcome. Robert Wood Johnson Foundation. http://www.activelivingresearch.org/files/ALR_Brief_ActiveTransport.pdf. Accessed Sept 2010
2. Alliance for a Healthier Generation (n.d.) Recess rocks/school wellness policies. http://www.healthiergeneration.org. Accessed 26 Sept 2010

3. Barros RM, Silver EJ, Stein RE (2009) School recess and group calssroom behavior. Pediatrics 123:431–436
4. Beighle A, Morgan CF, Le Masurier G, Pangrazi RP (2006) Children's physical activity during recess and outside of school. J School Health 76(10):516–520
5. Bungum TJ, Lounsbery M, Moonie S, Gast J (2009) Prevalance and correlates of walking and biking to school among adolescents. J Commun Health 34(2):129–134
6. Centers for Disease Control & Preventation (n.d.) Youth online: high school youth risk behavior survey: 1999–2009 [Data file]. http://www.cdc.gov/healthyyouth/yrbs/pdf/us_summary_all_trend_yrbs.pdf. Accessed Sept 2010
7. Centers for Disease Control & Prevention (2005) Barriers to children walking to or from shool—United States, 2004. MMWR [serial online] 54:949–952. http://www.cdc.gov/mmwr/preview/mmwrhtml/mm5438a2.htm. Accessed Sept 2010
8. Centers for Disease Control & Prevention (2010) The association between school-based physical activity, including physical education, and academic performance. US Department of Health and Human Services. http://www.cdc.gov/healthyyouth/health_and_academics/pdf/pa-pe_paper.pdf. Accessed Sept 2010
9. Chriqui JF, Schneider L, Chaloupka FJ, Ide K, Pugach O (2009) Assessing school district strategies for improving children's health. School Years 2006–07 and 2007–08. http://www.bridgingthegapresearch.org/_asset/hxbby9/WP_2009_monograph.pdf. Accessed Sept 2010
10. Chriqui JF, Schneider L, Chaloupka FJ, Gourdet C, Bruursema A, Ide K, Pugach O (2010) School district wellness policies: evaluating progress and potential for improving children's health three years after the federal mandate. School years 2006–07, 2007–08, and 2008–09, Vol. 2. http://www.bridgingthegapresearch.org/_asset/r08bgt/WP_2010_report.pdf. Accessed Sept 2010
11. Education Commission of the States (2005) Compulsory school age requirements. http://www.ecs.org. Accessed 26 Sept 2010
12. Frank LD, Sallis JF, Conway TL, Chapman JE, Saelens BE, Bachman W (2006) Many pathways from land use to health. J Am Plan Assoc 72(1):75–87
13. Hagan JF, Shaw JS, Duncan PM (2008) Bright futures: guidelines for health supervision of infants, children, and adolescents. Amercian Academy of Pediatrics, Elk Grove Village
14. Henderson CC (2003) The state of nutrition and physical activity in our schools. Environmental and Health. http://www.ehhi.org/reports/obesity/obesity_report04.pdf. Accessed Sept 2010
15. Jago R, Baranowski T (2004) Non-curricular approaches to increading physcial activity in youth: a review. Prev Med 39(1):157–163
16. Johnston LD, Delva J, O'Malley PM (2007) Sports participation and physical education in American Schools. Am J Prev Med 33(4):s195–s208
17. Center for Disease Control & Preventation. (n.d.). Components of coordinated school health. Retrieved 18 Sept 2010, from http://www.cdc.gov/healthyyouth/cshp/compoents.htm
18. Kann L, Brener ND, Wechsler H (2007) Overview and summary: school health policies and programs study 2006. J School Health 77(8):385–397
19. Kanters MA, Bocarro J, Casper J, Forrester S (2008) Determinants of sport participation in middle school children and the impact of intramual sports. RSJ 32(2):134–151
20. Lounsbery M, Bungum T, Smith N (2007) Physical activity opportunity in K-12 public school setting: Nevada. J Phys Act Health 4(1):30–38
21. McKenzie TL, Crespo NC, Baquero B, Elder JP (2010) Leisure-time physical activity in elementary schools: analysis of contextual conditions. J School Health 80(10):470–477
22. National Association for Sports and Physical Education (NASPE) (2004) Physical activity for children: a statement of guidelines for children ages 5–12. McGraw-Hill, Reston
23. National Association for Sports and Physical Education (NASPE) & Amercian Heart Asoication (2010) 2010 Shape of the Nation Report: status of physical education in the ISA. National Association for Sports and Physical Education. http://www.aahperd.org/naspe/publications/upload/Shape-of-the-tion-Revised2PDF.pdf. Accessed Sept 2010

24. National Center for Safe Routes to School (2010) Safe routes to school annual report 2008–2009. http://www.saferoutesinfo.org/about/annual_report_08-09/index.cfm. Accessed Sept 2010
25. National Institute of Child Health and Human Development Study of Early Child Care and Youth Development Network (2003) Frequency and intensity of activity of third-grade children in physical education. Arch Pediat Adol Med 157:185–190
26. National Federation of State High School Associations (NFSHHA) (2010) 2009–2010 High school athletics participation survery [Data file]. http://www.nfhs.org. Accessed Sept 2010
27. Parsad B., Lewis L (2006) Calories in, calories out: food and exercise in public elementary schools, 2005. National Center for Education Statistics, US Department of Education, Washington
28. Ridgers ND, Stratton, G, Fairclough SJ (2005) Assessing physical activity during recess using accelerometry. Prev Med 41(1):102–107
29. Robert Wood Johnson Foundation (RWJF) (2007, Sept). Recess rules. Sports4kids. http://www.playworksusa.org/files/rwjf_sports4kids_recess_report.pdf. Accessed Sept 2010
30. Robert Wood Johnson Foundation (RWJF) (2008, Feb). Overcoming obstacles to choosing health. http://www.rwjf.org/files/research/obstaclestohealth.pdf. Accessed Sept 2010
31. Robert Wood Johnson Foundation (RWJF) (2010, Feb). The state of play. http://www.rwjf.org/files/research/stateofplayrecessreportgallup.pdf. Accessed Sept 2010
32. Robert-Willson JE, Leatherdale ST, Wong SL (2008) Social-ecological corelates of active communting to school among high school students. J Adolesc Health 42(5):486–495
33. Rossell CH (1990) The carrot of the stick for school desegregation policy: magnet schools or forced busing. Temple University Press, Philadelphia
34. Sallis JF, Glanz K (2009) Physical activity and food environments: solutions to the obesity epidemic. Milbank Q 87(1):123–154
35. San Deigo State University, San Deigo California and Active Learning Research Program (2008, Jan) Physical education matters. The California endowment. http://www.calendow.org/uploadedFiles/Publications/By_Topic/Disparities/Obesity_and_Diabetes/PE%20Matters%20Long%20VersionFINAL.pdf. Accessed Sept 2010
36. SPARKfamily (2007) Math in motion activity unit. http://www.sparkfamily.org. Accessed 28 Nov 2007
37. Strong WB, Malina RM, Blimke CJ, Daniels SR, Dishman RK et al (2005) Evidence based physcial activity for school-age youth. J Pediatr 146(6):732–737
38. Sturgeon J, Meer J (2007) The first fifty years 1956–2006: the president's council on physical fitness and sports revisits its roots and charts its future. http://www.fitness.gov/home_pubs.htm. Accessed Sept 2010
39. Title IX (n.d.) History of title IX. http://titleix.info. Accessed Sept 2010
40. Trost S (2007) Active education: physical education, physical activity and academic performance (research brief). Active living research. Robert Wood Johnson Foundation, San Diego. http://www.activelivingresearch.com/alr/alr/files/Active_Ed.pdf. Accessed Sept 2010
41. Trost SG, Ward DS, Senso M (2010) Effects of child care policy and environmental on physical activity. Med Sci Sports Exerc 42(3):520–525
42. Tudor-Locke C, Lee SM, Pangrazi RP, Beighle A (2006) Children's pedometer-determined physical activity during the segmented school day. Med Sci Sports Exerc 38(10):1732–1738
43. Turner L, Chaloupka FJ, Chriqui JF, Sandoval A (2010) School policies and practices to improve health and prevent obesity: National Elementary School Survey Results: school years 2006–07 and 2007–08. University of Illinois at Chicago Bridging the Gap Program, Health Policy Center, Insitute for Health Research and Policy, Chicago. http://www.bridgingthegapresearch.org. Accessed 2010
44. US Department of Education (1997, June) Title IX: 25 years of progress. http://www2.ed.gov/pubs/TitleIX/index.html. Accessed 26 Sept 2010
45. US Department of Education, National Center for Education Statistics, Fast Response Survey System (FRSS) (2005) "Foods and Physical Activity in Public Elementary Schools: 2005," FRSS87. http://nces.ed.gov/pubsearch/pubsinfo.asp?pubid=2006106. Accessed 21 Aug 2010

46. US Department of Transportation (1972) The 1969 Nationwide personal transportations Study. US Department of Transportation Federal Highway Administration. http://www.fhwa.dot.gov/ohim/1969/1969page.htm. Accessed 26 Sept 2010
47. US Department of Transportation (2004) 2001 National Household Travel Survey [Data file]. US Department of Transportation Ferderal Highway Administration. http://nhts.ornl.gov/publications.shtml. Accessed Sept 2010
48. US Department of Health and Human Services (n.d.) Getting American moving: the history of the president's council on physical fitness and sports—1953–2002. http://www.fitness.gov/history.pdf. Accessed 26 Sept 2010
49. Willenberg LJ, Ashbolt R, Gibbs L, Garrard J, Green J B, Waters E (2010) Increasing school playground physical activity: a mixed methods study combining environmental measures and children's perspectives. J Sci Med Sport 13(2):210–216
50. Women's Sport Foundation (2005) Title IX Q & A. http://www.womenssportfoundation.org. Accessed 21 Aug 2010

Chapter 14
Primordial Prevention Through School Health Promotion

Gerald S. Berenson and Sandra Owen

Abstract The widespread occurrence of adult cardiovascular diseases in westernized populations, coronary artery disease, hypertension and diabetes mellitus, and the evidence these begin in childhood show the need to begin prevention early for virtually everybody. Damaging lifestyles learned early have to be addressed through education. The background from pediatric epidemiologic studies over the past four decades stresses approaches for prevention should be through primordial prevention. This can best be done through comprehensive and coordinated health education of school children. However, the challenge remains with respect to incorporating a sufficient health education program into the school systems that serve the entire population of children. Our experience shows single focused prevention efforts, e.g. for smoking, or for drugs, or for alcohol, are not as effective as broad health education, just like medical care is the ideal approach that physicians try to use in general medical practice. Thus, there is a need to get the parents and education community along with the communities of business and health care involved. Health education or prevention programs spearheaded by programs outside of the traditional educational system fail once the funding of such programs disappears. Failures also occur because general health is not part of the traditional education process and teachers are not trained in this discipline. Federal, state, and local governments will have to support school systems to begin and sustain such programs over time. Meeting this challenge is the best hope for primordial prevention.

Keywords Primordial prevention • Health education • Elementary schools • Community support • Family involvement

---

G. S. Berenson (✉)
Department of Medicine, Pediatrics, Biochemistry, Epidemiology, Center for Cardiovascular Health, Tulane University School of Medicine and School of Public Health and Tropical Medicine, New Orleans, LA, USA
e-mail: berenson@tulane.edu

G. S. Berenson (ed.), *Evolution of Cardio-Metabolic Risk from Birth to Middle Age*, 199
DOI 10.1007/978-94-007-1451-9_14, © Springer Science+Business Media B.V. 2011

14.1 Background

Primordial prevention aims to impede the penetrance of risk factors into a specific population by intervening in community settings that shape wellbeing; by targeting behavioral change, specific risk factors including those for heart disease; and by creating policy that promotes health and wellbeing [1]. Preventing the development of cardiovascular risk factors at the childhood age has the maximum potential to prevent or delay cardiovascular disease in adults and enhance longevity. Several observations made in the Bogalusa Heart Study illustrate family (adult-childhood) risk factor control has a favorable impact on reducing cardiovascular disease and related longevity (albeit indirect evidence).

The Bogalusa Heart Study has demonstrated that, although cardiovascular disease in terms of anatomic changes begin in childhood by 5–8 years of age, children with a few or no risk factors had significantly less carotid intima-media thickness, a measure of subclinical atherosclerosis as adults, Fig. 14.1. Also, we documented a significant relationship to metabolic syndrome in adults with and without a family history of longevity (lived after 85 years of age), cardiovascular disease, hypertension, and type 2 diabetes, Fig. 14.2 [2]. Also, the prevalence of risk factor clustering occurring at the lowest quartile related to a negative family history of heart disease and other chronic diseases, Fig. 14.3. These findings implicate the potential beneficial impact of preventing the development and controlling risk factors in early life.

14.2 Approach to Prevention

The most promising health education goes beyond the cognitive level and addresses health determinants, social factors, attitudes, values, norms, and skills shown to influence specific health-related behaviors [3–5]. Unfortunately, early adoption of poor habits by children learned at home, and those practiced under pressure from peers, hinders the child's ability to mature into a healthy adult. Social factors like teen pregnancy, drug induced depression, and antisocial/violent behavior trigger a child's academic failure and subsequently cause him/her to drop out of school.

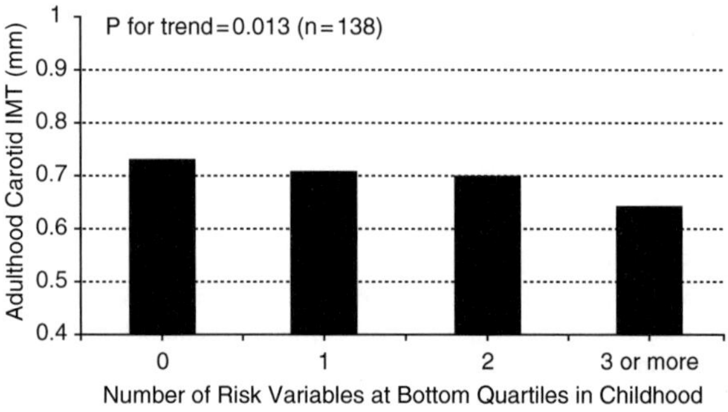

Fig. 14.1 Carotid IMT measured in adults by the number of risk variables at the bottom quartiles in their childhood. Carotid IMT is the average of common, internal, and bulb segments (three *left* and three *right* far wall measurements). [2]

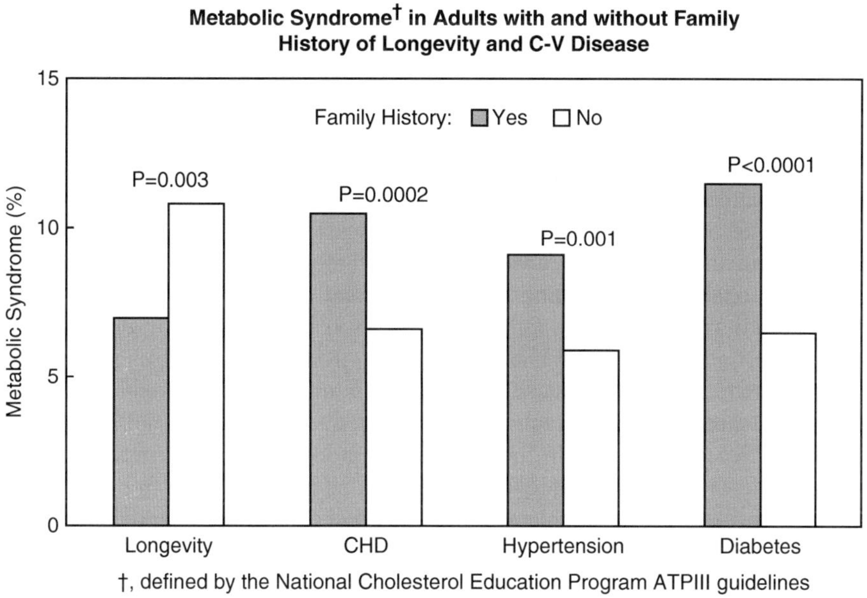

Fig. 14.2 Metabolic syndrome in adults with and without family history of longevity and C-V disease

Fig. 14.3 Parental history of CV disease and prevalence of clustering of three or more risk variables at the bottom quartile in childhood. A positive family history was defined as mother and/or father having a history of coronary heart disease, hypertension or type 2 diabetes. [2]

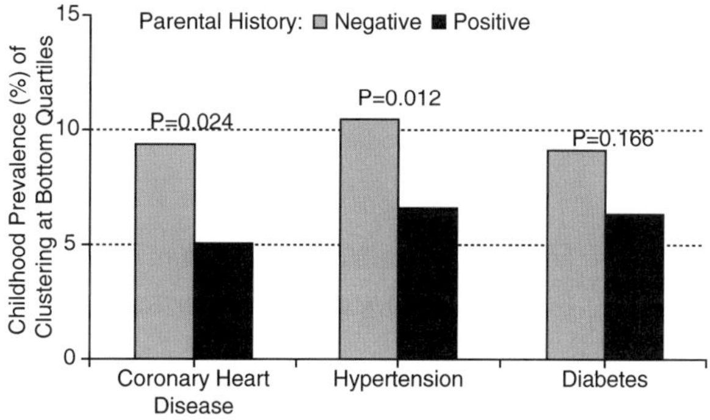

In some states, the drop-out rate approaches 35–40%. Primordial school-based interventions are a challenge to teach components of health literacy (effective communication, decision-making, problem-solving, self-initiated learning, and personal responsibility for ones behavior) in order to empower children and promote their sense of self-worth/self-esteem. Learning experiences that promote self-efficacy very early in the child's education career prepare children to be responsible for their health behavior and reduce the likelihood of practicing risk behaviors.

Comprehensive and coordinated health programs are to be adopted that address the entire school environment [6]. Education broadly on health, needs to achieve an emphasis equal to that given to traditional subjects—English, Math, etc. Obviously, this is not happening—teachers are not trained in broad aspects of health as we propose, ideal classroom materials are limited, and time and pressure for teaching are directed to specific and traditional educational areas. Yet, education promoting healthy lifestyles and healthy decision-making can begin in preschool and kindergarten and extend throughout the elementary school age. This age captures an important learning period of life to begin to adopt healthy lifestyles and encouraging children not to undertake unhealthy lifestyles. Such a background sets the stage for lifelong health and lifestyles that will prevent or delay heart disease, obesity and diabetes. Importantly, it will also improve student performance, through improved social behavior.

The rationale for introducing a health program is that good health promotes sound minds as a prerequisite for a good education system. Unfortunately, misbehavior, early adoption of poor habits from home, from peers and at school pervade early life and thwart good education habits and interferes with prevention of adult chronic diseases, our goal. Teaching self-esteem, decision making and incorporating self-efficacy very early in the educational process are part of teaching very young students to be responsible for their health and inoculate them from undertaking bad lifestyles. Social problems like violent behavior, teen age pregnancy and drop-outs must be addressed. Parents, teachers, and community have to be involved to support health promotion and make it successful [7, 8].

There is a growing awareness for the need of health promotion in schools brought on by the epidemic of obesity and increasing type II diabetes in children and adolescents. So it is possible schools will gradually undertake introducing broad health education promotion like Health Ahead/Heart Smart, including physician involvement [9–11]. Unfortunately, many health education efforts target only one behavior, such as smoking or exercise and now obesity. We note single focused efforts are not effective. A global approach must involve the classroom curriculum, nutrition on and off the school grounds, physical education, school health services and in-service education.

14.2.1 Health Ahead/Heart Smart K-6 School Based, Primordial Prevention

As an example a comprehensive and tested program like "Health Ahead/Heart Smart" (K-6) helps with such problems [9–11]. It is the concept of public health education of school children that needs to be adopted and implemented with Pediatricians and Cardiologist providing leadership and guidance [12].

Health Ahead/Heart Smart K-6 curriculum integrates national health education, science, and technology standards for which schools are currently held accountable. It models comprehensive and coordinated school health education and incorporates school guidelines for nutrition, physical activity, and tobacco use prevention recommended by the Centers for Disease Control and Health Promotion. Lessons include measurable objectives, valid assessments, technology resources, interactive student activities, and in/home application activities.

There are justifications for recommending comprehensive health education to become an integral part of the public education process:

1. "The curriculum and health promotion components address important issues such as nutrition, exercise, non-tobacco use, and alcohol and drug abuse, violent behavior and prevention of infectious diseases.
2. The curriculum and health promotion components address important issues of social concern. By encouraging students to adopt healthy lifestyles and coping skills, the program helps students mature, take care of themselves and be responsible for others.
3. A program of health promotion which motivates children to take an interest in their own well-being encourages a positive self-image and improved student involvement, potentially decreasing dropouts.
4. Improvement in lifestyles and health activities increases the capability for learning, thus improving student performance.
5. The health education program includes support mechanisms that also encourage health education in general. It improves the health and attitudes of the entire school environment, e.g. health fairs, improved school lunches and snack foods, exercise programs and involvement of teachers in the improvement of their own health.
6. The program involves parents and families, encouraging parental interest in children's learning and educational development as well as improved family health.
7. Comprehensive health education directed toward cardiovascular disease, cancer, obesity and social issues related to disease can improve the health of our nation".

The above is taken from "Introduction of Comprehensive Health Promotion for Elementary schools" [6].

The Health Ahead/Heart Smart is unique in its strong behavioral orientation including mastery experiences, empowerment activities, and parent and community outreach to achieve healthy lifestyles. Its comprehensive curriculum includes exercises in decision- making and problem-solving specific to avoiding tobacco and alcohol use; personal nutrition and physical activity assessments; and writing activities about the importance of valuing education and staying in school. There are activities suggested to involve and gain interest by teachers, Table 14.1.

Four modules are contained in each grade level:

(1) health behavior, general health and cardiovascular risk (anatomy and physiology); (2) nutrition; (3) exercise; and (4) psychosocial skills, including assertiveness

Table 14.1 Suggested activities to involve parents and teachers. [18]

- Kick-off event
- Health week (Heart Smart Week)
- Health fair
- Newsletter (including recipe modifications, new recipes, health information)
- Afternoon Lifestyle Program for Teachers (i.e., walking club, menu planning)
- Role Model Program (parents, business leaders, professionals, athletes)
- Nutrition club (cafeteria manager and upper-grade students)
- Upper-grade students teaching in lower grade
- Big Buddy Program to teach safety (i.e., wearing seat belts, helmets)
- Geographic Fun Run
- Planting club to grow garden of fresh vegetables

and coping strategies. In grades K-3, a fifth module called "It's me" emphasizes self-identity, a positive self-image, the importance of taking responsibility for ones health, and inoculation against adoption of unhealthy lifestyles. An emphasis on social learning through cross-age, peer teaching is an important part of each grade level curriculum. Additional components of the program include "Gimme Five" (nutrition) [13], "Super Kids/Super Fit" (physical activity) [14], a school lunch room manager training manual, and a family health program [14–16].

Qualitative feedback from teachers included lunchroom observations of students adding less to no salt on their food and more often selecting salads, whole wheat bread, and fruit in place of sweet desserts. The lunchroom manager removed fries from the menu. Students' self-report pre-post intervention documented a 26% increase in students who think exercise is important; a 10% increase in students eating breakfast; a 9% increase in students exercising 3–5 days a week; a 4% increase in students working to lose weight; and a 3% increase in students setting personal goals [17].

14.2.2 Exercise and Wellness

With the crisis of the obesity epidemic and knowledge that children today are less physically fit, it is crucial to provide specific programs to address physical activity. One such program is "Superkids/Superfit" [14]. An exercise curriculum is a crucial part of the comprehensive and coordinated "Health Ahead/Heart Smart" health education program (K-6). "Superkids/Superfit" was designed by Stephen Virgilio, PhD, and updated by two graduate students, Jackie Epping, MEd, and Patrice Strikemiller, MS.

A guidebook or physical activity curriculum is available and a menu box of activity cards with approximately 75 exercises including noncompetitive activities. The variety of exercises enables addressing all children, including those who are obese, and not just athletes. One evaluation can be done by a run-walk one mile test or by incorporating repeated performance to the President's Fitness Challenge with a ¼ mile run and other activities. The exercise component has been shown to be capable of controlling obesity and improving exercise performance [18].

The primary purpose, along with exercise, is to develop knowledge, behaviors and attitudes related to physical activity which are consistent with lifelong maintenance of cardiovascular health. Active participation of all students is encouraged and regular physical activity outside of school's physical education class is promoted. The program recognizes limited space and limited budget. Therefore activities are provided that can be implemented by classroom teachers as well as physical education teachers. Some schools do not have outdoor facilities for exercise and inclement weather may preclude exercising outdoors, therefore activities contained in the menu box can serve for classroom exercise. This concept was taken from the "Feeling Good" program by Kuntzleman [19]. The role models, teachers and parents, are encouraged to improve their own health lifestyles through participating in some of the physical activities, e.g. run-walk at the time of health fairs or at other times.

Adaptation of the "Superkids/Superfit" exercise component was studied by Mott and colleagues [20] as personalized fitness modules including a knowledge test and the children's Attitudes Towards Physical Activity Inventory. The one-mile run/walk was studied to determine change in performance. The personalized unit substantially improved the one-mile run/walk time indicating use of such modules in health and physical education programs may enhance life-long exercise and fitness.

14.2.3 Cardiovascular Risk Factor Screening

The extent of screening of faculty, staff, students and parents depends on resources available. Screening of all four groups promotes self-responsibility for health promotion behavior. It helps reflect curriculum topics and provides a positive climate for role rehearsal and modeling of desired health behaviors.

With limited resources, height, weight, body mass index (BMI) and blood pressure can be obtained on each child. A Family History is useful and a personal history of vaccinations and illnesses should be obtained. If possible, detailed screening for a lipid profile is recommended to be done by a physician interested in preventive cardiology. It is mandatory that each participating individual and the physician of his/her choice receive a report of their health issues and screening results. Each participating student should receive a report of their health issues and screening results to be recorded in a "Health Passport". The health record, like BMI and vaccinations, should follow the student's progression through subsequent grade levels.

14.2.4 Community Involvement

Community intervention strategies are used to support school health education and improve citizen's lifestyles. Washington Parish (county) community approach is an example of Health Ahead/Heart Smart reaching out to the community [18]. The program requires a large amount of volunteer help throughout the Parish and within the small communities to see activities implemented. A number of committees or task forces can be formed to help implement various activities. The purpose of each is briefly outlined.

14.2.4.1 Public Relations

This involves news-media, paper, radio, and television reporting activities to get the word out beyond Washington Parish, of what is being accomplished.

14.2.4.2 Medical, Dental and Nursing

Following major medical or other health problems of each child within the schools and referring or making available needed health care by doctors and Parish-wide facilities.

14.2.4.3 Business

Investigating how such a program can be financed and how fund raising through grant writing, for federal and state funds, can be accomplished.

14.2.4.4 Education

Assistance in implementing Health Ahead/HeartSmart K-6 curriculum throughout the Parish. This is to include all elementary school children in the Parish including the small rural communities. The program addresses the entire school environment: nutrition, physical education, personal health programs for teachers and parents. It is important that the program also address social problems including smoking, alcohol, drugs, and violent behavior. It will require in-service training that may be accomplished by distant learning from a medical center or hospital. Parental involvement is encouraged.

14.2.4.5 Public Policy

Dialogue with mayors and their staff as well as the Parish (County) Sheriff. Thirty percent or more high school students are using tobacco products and drinking underage. Policy against the sale to minors needs to be enforced. Security guards will need to be provided to schools to enforce no use policy on campus grounds. Construction of bicycle paths and areas for planned and supervised exercise for the children in the community will have to be discussed.

14.2.4.6 Evaluation

The formation and implementation of a formative and summative program evaluation plan uses quantitative and qualitative data. Process evaluation of committee activities and accomplishments is of special interest.

14.3 Summary

Health problems like cardiovascular diseases and unhealthy lifestyles begin in childhood. School-based primordial prevention promotes early risk prevention or reduction and health promoting behaviors. Such an approach must be comprehensive and coordinated involving parents and the community at large. To be successful, schools must continually assess policy, curriculum, and teacher preparedness. They must first understand and appreciate the influence of health on student performance and achievement and secondly, consider health education equally important as math, science, etc. Interactive learning activities must be visible in the curriculum if students are to be empowered to be responsible for their own health behavior. As the school reaches out to the community, a "Healing Community" is created in which local resources and manpower are pulled together to create an independent, self-sustaining capability required to maintain primordial prevention of children.

References

1. American Heart Association Scientific Statement (2003) AHA Guide for improving cardiovascular health at the community level. Circulation 107:645–651
2. Chen W, Srinivasan SR, Li S, Xu J, Berenson GS (2005) Metabolic syndrome variables at low levels in childhood are beneficially associated with adulthood cardiovascular risk: the Bogalusa Heart Study. Diabetes Care 28:126–131
3. National Center for Chronic Disease Prevention and Health Promotion (2008) Characteristics of an effective education curriculum. School Health Education Resources (November 2008)
4. Brenner ND, McManus T, Foti K, Shanklin SL, Hawkins J, Kann L, Speicher N (2008) Characteristics of health programs among secondary schools. School Health Profile 2008. Centers for Disease Control and Health Promotion, Atlanta
5. Owen S (1980) When the student is ready the teacher will come: a perspective on health education. J S Carolina Med Assoc 76:16–20
6. Berenson GS et al (1998) Introduction of comprehensive health promotion for elementary schools: the Health Ahead/Heart Smart Program. Vantage Press, New York
7. Nader PR, Perry C, Maccoby N, Solomon D, Killen J, Telch M, Alexander JK (1982) Adolescent perceptions of family health behavior: a tenth grade educational activity to increase family awareness of a community cardiovascular risk reduction program. J Sch Health 52:32–37
8. Downey AM, Greenberg JS, Virgilio SJ, Berenson GS (1989) Health promotion model for "Heart Smart": the medical school, university and community. Am J Health Promotion 13:31–46
9. Downey AM, Frank GC, Webber LS, Harsha DW, Virgilio SJ, Franklin FA, Berenson GS (1987) Implementation of "Heart Smart": a cardiovascular school health promotion program. J Sch Health 57:98–104
10. Downey AM, Cresanta JL, Berenson GS (1989) Cardiovascular health promotion: "Heart Smart" and the changing role of the physician. Am J Prev Med 5:279–295
11. Downey AM, Virgilio SJ, Serpas DC, Nicklas TA, Arbeit ML, Berenson GS (1988) "Heart Smart"—a staff development model for a school-based cardiovascular health intervention. Health Educ 19:64–71
12. Berenson GS, Srinivasan SR, Fernandez C, Chen W, Xu J (2010) Can adult cardiologist play a role in prevention of heart disease beginning in childhood? Methodist Debakey Cardiovasc J 6:4–9
13. Nicklas TA, Johnson CC, Farris R, Rice R, Lyon L, Shi R (1997) Development of a school-based nutrition intervention for high school students: Gimme 5. Am J Health Promot 11:315–322
14. Virgilio SJ, Berenson GS (1988) Superkids-Superfit: a comprehensive fitness intervention for elementary schools. J Phys Educ 59:19–25
15. Johnson CC, Nicklas TA, Arbeit ML, Harsha DW, Mott DS, Hunter SM, Wattigney W, Berenson GS (1991) Cardiovascular intervention for high-risk families: the Heart Smart Program. South Med J 84:1305–1312
16. Berenson, GS, Harsha DW, Johnson CC, Nicklas TA (1993) Teach families to be heart smart. Patient Care 6:135–145

17. Arbeit ML, Johnson CC, Mott DS et al (1992) The Heart Smart cardiovascular school health promotion. Behavior correlates of risk factor change. Prev Med 21:18–32
18. Berenson GS (2010) Cardiovascular health promotion for children: a model for a Parish (county)—wide program (implementation and preliminary results). Prev Cardiol 13:23–28
19. Kuntzleman CT (1985) Instructors guide for feelin' good. Fitness Finders, Spring Arbor
20. Mott DS, Virgilio SJ, Warren BL, Berenson GS (1991) Effectiveness of a personalized fitness module on knowledge, attitude and cardiovascular endurance of fifth-grade students: "Heart Smart". Percept Mot Skills 73:847–858

Index

G. S. Berenson (ed.), *Evolution of Cardio-Metabolic Risk from Birth to Middle Age*, DOI 10.1007/978-94-007-1451-9, © Springer Science+Business Media B.V. 2011